# Optical Wideband Transmission Systems

Editor

## Clemens Baack, Dr.-Ing.

**Professor**
Heinrich-Hertz-Institut für Nachrichtentechnik
Berlin, West Germany

**CRC Press, Inc.**
**Boca Raton, Florida**

Library of Congress Cataloging in Publication Data
Main entry under title:

Optical wideband transmission systems.
    Bibliography: p.
    Includes index.
    1. Optical communications. 2. Laser communication
systems. 3. Fiber optics. I. Baack, Clemens, 1937-
TK5103.59.0687    1986       621.38'0414       85-4244
ISBN 0-8493-6152-4

Direct all inquiries to CRC Press, Inc., 2000 Corporate Blvd., N.W., Boca Raton, Florida, 33431.

© 1986 by CRC Press, Inc.

International Standard Book Number 0-8493-6152-4
Library of Congress Card Number 85-4244
Printed in the United States

# THE EDITOR

Clemens Baack, Ph.D. received the Ing. degree in electrical engineering in 1959 and the Dipl.-Ing. degree in communication engineering in 1967 from the Rheinische Ingenieurschule Bingen and the Technische Universität Berlin, respectively. From 1968 to 1970 he was with the Hahn Meitner Institut für Kernforschung in Berlin, where he was engaged in the development of electronic circuits. From 1970 to 1974 he was research associate at the Institut für Hochfrequenztechnik of the Technische Universität Berlin, where he received his Dr.-Ing. on Phase-Array-Antennas-Theory. From 1974 to 1975 he was with the Forschungsinstitut für Funk und Mathematik, Wachtberg-Werthoven, where he worked in the field of Phased-Array-Radar-Systems.

In 1975, Dr. Baack joined the Heinrich-Hertz-Institut für Nachrichtentechnik, where he headed the group which was engaged in broad-band transmission techniques. In 1980 he became the head of the Department Switching and Transmission of the Institut.

In 1982, he was appointed as Professor for Wideband Communications at the Technische Universität Berlin and as Scientific Director of the Heinrich-Hertz-Institut.

# CONTRIBUTORS

Wolfgang Albrecht
Dr.-Ing.
Heinrich-Hertz-Institut für
  Nachrichtentechnik
Berlin, West Germany

Clemens Baack
Professor
Heinrich-Hertz-Institut für
  Nachrichtentechnik
Berlin, West Germany

Gerhard Elze
Dipl. -Ing.
Heinrich-Hertz-Institut für
  Nachrichtentechnik
Berlin, West Germany

Bernhard Enning
Dipl.-Ing.
Heinrich-Hertz-Institut für
  Nachrichtentechnik
Berlin, West Germany

Gerd H. Grosskopf
Dr.-Ing.
Heinrich-Hertz-Institut für
  Nachrichtentechnik
Berlin, West Germany

Günter Heydt
Dipl.-Ing.
Heinrich-Hertz-Institut für
  Nachrichtentechnik
Berlin, West Germany

Lutz Ihlenburg
Dipl.-Ing.
Heinrich-Hertz-Institut für
  Nachrichtentechnik
Berlin, West Germany

Peter Meissner
Dr.-Ing.
Heinrich-Hertz-Institut für
  Nachrichtentechnik
Berlin, West Germany

Heribert Münch
Dipl.-Int.
Heinrich-Hertz-Institut für
  Nachrichtentechnik
Berlin, West Germany

Erwin Patzak
Dr. rer. nat.
Heinrich-Hertz-Institut für
  Nachrichtentechnik
Berlin, West Germany

Godehard Walf
Dipl.-Ing.
Heinrich-Hertz-Institut für
  Nachrichtentechnik
Berlin, West Germany

Gerhard Wenke
Dipl.-Phys.
Heinrich-Hertz-Institut für
  Nachrichtentechnik
Berlin, West Germany

# TABLE OF CONTENTS

Chapter 1
Introduction.................................................................................. 1
Clemens Baack and Gerhard Elze

Chapter 2
General System Considerations................................................ 3
Clemens Baack, Gerhard Elze, Lutz Ihlenburg, and Heribert Münch

Chapter 3
Single-Mode Fibers.....................................................................11
Gerhard Elze, Heribert Münch, and Gerhard Wenke

Chapter 4
Optical Sources for Broadband Transmission Systems ......................21
Gerd H. Grosskopf, Peter Meissner, Erwin Patzak, Godehard Walf, and
Gerhard Wenke

Chapter 5
Coupling Efficiency and Optical Feedback Characteristics...................55
Gerhard Elze and Gerhard Wenke

Chapter 6
Optical Detectors .......................................................................73
Wolfgang Albrecht, Gerhard Elze, and Gerd H. Grosskopf

Chapter 7
Aspects of Broadband Circuit Design..........................................87
Wolfgang Albrecht, Bernhard Enning, and Godehard Walf

Chapter 8
Optoelectronic Receivers.......................................................... 119
Wolfgang Albrecht and Clemens Baack

Chapter 9
Signal Generation and Regeneration ......................................... 147
Bernhard Enning, Günter Heydt, Lutz Ihlenburg, and Godehard Walf

Chapter 10
Remarks on System Performance............................................. 183
Clemens Baack, Gerhard Elze, Bernhard Enning, and Gerhard Wenke

Chapter 11
Future Aspects........................................................................ 195
Clemens Baack, Gerhard Elze, Günter Heydt, and Gerhard Wenke

Index...................................................................................... 205

Chapter 1

# INTRODUCTION

## Clemens Baack and Gerhard Elze

The progress made in microelectronics has made possible the development of integrated services digital networks (ISDN) which can be used to supplement the telephone service by offering numerous additional text, data, and facsimile services. Optical fibers are the most suitable medium for the transmission of the large amounts of information that can be expected for local trunk and particularly long-distance trunk operations. As a result of the wide repeater spacing that can be achieved, optical communications technology permits the transmission of large quantities of digitalized information over considerable distances at a reasonable cost. The projected upgrading of ISDNs into broadband ISDNs (BB-ISDN) in many countries is likely to lead to a further significant increase in the volume of information at trunk levels in networks of the future. In particular, video telephones will impose high demands at trunk level.

The bit rates to be used in future digital communication networks are determined by the PCM (pulse code modulation) hierarchy in the various regions. Already today optical transmission systems in the lower and middle network levels with bit rates of up to several hundred megabits per second are deployed in the field. Optical transmission systems in the highest levels of the PCM hierarchies in the 1 to 2 Gb/sec range have already progressed beyond the laboratory stage in several countries and are being used in field trials. This book is concerned with these high bit rate systems, which are predestined particularly for the long-distance trunk lines that will be used in future communications networks.

For a number of years the authors have been involved with digital and analogue optical BB transmission technology, through their work on various research projects. The experience thus obtained with digital gigabits per second (Gb/sec) systems serves as the basis for this book. An effort has also been made to give readers a wider overview of the work currently being carried out in this area, and to illustrate the problems which still remain to be solved.

Serving as a point of reference for all the subsequent chapters of this book, Chapter 2 describes the structure and elements of an optical, digital transmission system. Single-mode (SM) fibers are the only suitable transmission medium for such high-rate systems, and graded-index fibers must be ruled out. Due to mode dispersion, the bandwidth distance product of graded-index fibers is too small. In addition, intolerable restrictions on the bit rate-distance product occur as a result of interference in multimode (MM) fibers, such as modal noise effects. The systems under discussion will be operated in the optical long-wavelength range ($\lambda = 1.3$ to $1.6\ \mu m$) where SM fibers offer miminal total dispersion with a minimum of attenuation, and thus an almost unlimited transmission capacity.

Further on, Chapter 2 gives the maximum distance attainable between repeaters that can be achieved in the optical long-wavelength range using SM fibers. For these calculations the noise produced by the optoelectronic receivers (OER), the attenuation and dispersion of the fiber, and the extinction ratio of the laser must be taken into account. Other additional interference sources, e.g., those resulting from the MM behavior of the laser, which are of particular importance to high-rate systems, have not been taken into account in this chapter. These problems are described in Chapter 4 and Chapter 10.

Chapter 3 provides a brief summary of the main transmission characteristics of SM fibers and of the processes known at the present time to compensate the dispersion.

Chapter 4 lists the demands that are imposed on semiconductor lasers intended for use in optical systems with high bit rate-distance products. In order to avoid the additional interference sources already mentioned, efforts should be made to use lasers with dynamic single-mode behavior (DSM lasers). One of the main objectives of this chapter is to acquaint the reader with the various processes known today for stabilizing laser spectra and for suppressing side modes — and to compare these various processes in terms of their suitability for their application in BB systems.

Lasers, SM fibers, and photodiodes must be coupled together as effectively as possible, using practicable methods. Chapter 5 provides an outline of the basic details for SM fiber coupling. Not only does coupling efficiency play a decisive role in high-rate systems, but in addition, the light components reflected from various coupling points are also a significant factor affecting DSM behavior of the lasers. The laser-fiber coupling is examined in detail with regard to the degree of efficiency of the coupling, and to the optical reflection.

Chapter 6 gives a brief summary of photo detectors and their important characteristics which are necessary to consider when designing OERs. The high-rate light signals converted by the photodiode must be regenerated by means of electronic circuits. Currently, hybrid circuits manufactured in thick or thin film technology are mainly used for processing digital signals in the 1 to 2 Gb/sec range. However, in the future one can expect a changeover to monolithic circuits on a Si or GaAs basis. To keep design costs for hybrid and particularly for monolithic circuits within reasonable limits, the use of powerful network analysis programs is essential. As shown in Chapter 7, it is necessary to characterize the various passive components and their parasitic elements precisely. This also involves very complex equivalent circuit diagrams. The results summarized in this chapter are intended as an aid in the design of the circuits dealt with in subsequent chapters.

The bit rate-distance product of an optical transmission system is determined to a decisive extent by the sensitivity of the OERs, i.e., the combination of a photodiode followed by an amplifier.

A number of circuit concepts for use in setting up extremely BB-OERs are analyzed in considerable detail in Chapter 8. The various BB amplifiers are combined with different photo detectors, in order to be able to compare the attainable receiver sensitivities as a function of the bit rate. The general considerations are then verified by implemented OERs in the 1 and 2 Gb/sec range.

Pulse shaping must be carried out in a number of different ways in transmitters and receivers of a digital transmission system. Some of the circuit concepts proved effective for low-rate systems are ineffective for the design of high-rate systems. Other concepts had to be investigated, e.g., the use of delay lines in circuits for Gb/sec applications. Chapter 9 describes circuits for pulse shaping processes which have proven effective in realized systems. Another main aspect covered by this chapter concerns the problem of clock regeneration in general, as well as the design and implementation of circuits which are suitable for use in Gb/sec systems.

The main findings from these chapters have been implemented and tried out as part of a complete high-rate experimental system: in accordance with the highest stage of the European PCM hierarchy currently under discussion, a 2.24 Gb/sec system was implemented for a light wavelength of approximately 1.3 $\mu$m. Chapter 10 displays the results of the measurement of system data and describes the disturbances for optical transmission lines.

Finally, Chapter 11 seeks to outline feasible development trends in optical BB communication. The future of optical BB transmission technology will be determined to a decisive extent by developments in electronic high-speed integrated circuits (HSIC) and by the development of optoelectronic integrated circuits (OEIC) as well as by the possibilities offered by optical heterodyne technology.

Chapter 2

# GENERAL SYSTEM CONSIDERATIONS

Clemens Baack, Gerhard Elze, Lutz Ihlenburg, and Heribert Münch

## TABLE OF CONTENTS

I.  Structure and Components of a Digital, Optical Broadband (BB)
    Transmission System..................................................................................4

II. Digital, Optical BB Transmission Taking Into Account the Theoretical
    Aspcets of the System.................................................................................5
    A.    Conditions for Calculation .......................................................5
    B.    Results..........................................................................................7

References .................................................................................................10

## I. STRUCTURE AND COMPONENTS OF A DIGITAL, OPTICAL BROADBAND (BB) TRANSMISSION SYSTEM

Figure 1 shows the structure of the digital, optical transmission system consisting of the digital source, a transmitter filter for preprocessing of the signal, the optical transmission channel with subsequent signal regeneration, and digital sink.

Adjustment of the digital source signal to the properties and transmission behavior of the optical transmission channel takes place by means of selected pulse shaping in the transmitter filter. The output signal of the transmitter filter is the transmission pulse sequence, the individual pulses of which are adjusted in both form and duration to the properties of the transmission channel, to ensure that transmission is as error-free as possible.

A laser diode is used to produce the optical signal (Chapter 4). It is essential to avoid modal dispersion and modal noise to achieve highest transmission rates with maximum repeater spacing for a given bit error rate; therefore single-mode (SM) fibers must be used. The use of SM fibers in conjunction with lasers with a broad spectrum, i.e., longitudinal, multimode lasers, produces reduction in bandwidth due to chromatic dispersion and laser-mode-partition-noise (LMPN; see Chapter 4). These perturbations lead to a significant reduction in repeater spacing. LMPN can be avoided by using a laser with a monomodal spectrum. Under this condition, the resultant broadening of the pulses due to chromatic dispersion within the fiber is very small. The destructive effects of the spectral behavior and of noise produced as a result of optical feedback can be suppressed by means of laser mode selection, i.e., by means of an external resonator used in conjunction with an optical isolator (although this does cause additional attenuation, a factor which should also be taken into consideration; see Chapters 5 and 10).

If direct modulation of the laser is used, fluctuations in carrier density and temperature in the active zone will interfere with the spectral behavior. However, this can be avoided by applying external optical modulation.

The optical transmission signal, processed as required in order to obtain transmission as error-free as possible, is fed into the transmission medium using the appropriate coupling elements (Chapter 5).

Transmission in the optical long wavelength range ($\lambda = 1.3$ to $1.6\ \mu m$) is technologically feasible today, and it is now possible to operate systems in which there is low ($\lambda = 1.3\ \mu m$) or minimal fiber attenuation ($\lambda = 1.55\ \mu m$, Quartz) and relatively low residual dispersion (Chapter 3), thus permitting a considerable increase in the bit rate-distance product attainable.

Due to perturbations along the fiber transmission line (Chapter 3), additional signal-dependent and -independent noise components are introduced at the photodiode, which converts the output signal of the fiber (Chapter 6) to an electrical signal.

Signal amplification by means of an amplifier with a low-noise, front-end circuit (Chapter 8) is followed by the receiver filter (Chapter 9). It is expedient to make a distinction between equalizer and filter stages within the receiver filter. The task of the equalizer circuit is to equalize the transmission channel, whereas the filter serves to produce a pulse sequence which exhibits the characteristic properties of the chosen transmission method, i.e., Nyquist or Gaussian pulses (see Section II) and to restrict bandwidth.

A device for amplitude and time regeneration, consisting of a 1 or $\mu$-1 level comparator and a sampler is used to perform the regeneration of the digital information. The sampling device requires the timing information produced in the clock regeneration circuit. The use of quantized feedback equalization (QFE) in order to reduce the pulse interferences, by compensating the postcursor (low-pass QFE), without affecting detection noise performance, is an optional feature and depends on the transmission procedures chosen (see Section II). QFE for regeneration of the base line (high-pass QFE) is necessary (without special coding) because of the capacitative couplings in the receiver.

## II. DIGITAL, OPTICAL BB TRANSMISSION TAKING INTO ACCOUNT THE THEORETICAL ASPECTS OF THE SYSTEM

Using the calculation methods shown in References 1 through 5, this chapter will compare various transmission methods for optical BB transmission systems. A prerequisite is that such systems are attenuation-limited, i.e., systems with SM fibers and dynamic single-mode (DSM) lasers, the maximum repeater spacing of which is determined solely by the fiber attenuation and not by the fiber bandwidth.

These calculations are based on the following system parameters: laser output power, extinction ratio of the laser output pulses, spectral linewidth of the laser, attenuation and dispersion of the fiber, and the noise properties of the photodiode and the electronic amplifier. Thus, the calculated repeater spacings are optimized values, which are not attainable in practice at present. The numerous additional sources of interference, which are particularly pronounced in the case of high-rate systems, and which are due to the interaction of the laser with the fiber, are not taken into account here. These interference sources and their suppression are subjects of world-wide interest, and more attention will be paid to them in subsequent sections.

A comparison will be made between binary and multilevel systems. For comparison purposes, Nyquist and Gaussian pulses are produced by the receiving filter at the input to the clocked comparator (Figure 1).[6,8] Finally, in the case of Gaussian pulses, the influence of the low-pass QFE is investigated for the suppression of interference of the postcursor.

### A. Conditions for Calculation

The DSM transmitter module (Figure 1) has a maximum optical power of 3.5 mW with a spectral linewidth (full width at half maximum) of 0.1 nm (Chapter 4). The maximum optical power resulting from the bias current of the laser diode is assumed to be 0.5 mW (Chapter 4). The attenuation produced by the laser-fiber coupling is 3 dB (Chapter 5).

It is already known that the highest signal-to-noise ratio (SNR) in a digital transmission system is achieved at the input to the comparator by means of joint optimization of the transmission pulse form by use of the transmitter filter and the receiving pulse form by the receiver filter. The optimization rules are well known for white, additive noise, e.g., for copper cable systems.[9] These rules can be easily extended to cover colored noise too. The authors are not aware of optimization rules for the optical systems under consideraton here, with additional signal-dependent noise in the photodiode. Baack[10] shows that the results for rectangular transmission pulses (theoretical maximum edge steepness) and Gaussian-shaped transmission pulses (minimum edge steepness) hardly change at all. This is not surprising because as is known, the influence of the transmission pulse shape on the sensitivity is negligible as long as the significant component of the pulse energy is transmitted during the bit period. Since the possibilities for precise pulse shaping are limited anyway, in the high-rate systems under consideration here, no optimization of the transmitter filter is undertaken, and only the receiving filter is optimized. In close conformity to the conditions found in practice, it is assumed that the transmission pulse at the output of the transmitter filter is Gaussian-shaped. The maximum amplitude of the transmission pulse is obtained by means of the maximum laser output for DSM behavior. The width of the transmission pulse,

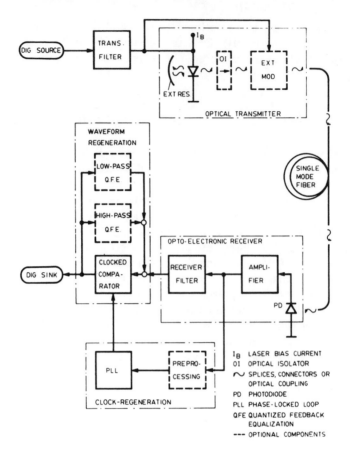

FIGURE 1.    Model of a digital optical BB transmission system.

which is defined by the distance between the points of inflection of the Gaussian curve, can be varied as an optimizable system value.

The transmission behavior of the SM fiber is assumed to be linear and to have a Gaussian-shaped impulse reponse.

Chapter 8 describes the selection of an optoelectronic receiver (OER) consisting of a GE-APD (avalanche photodiode) and a transimpedance amplifier for the 1 and 2 Gb/sec ranges under consideration here. The Ge-APD has the following properties:

- Quantum efficiency, 75%
- Dark current, 2 $\mu$A at an avalanche gain $M_d = 10$
- Excess noise exponent, 0.95

The noise behavior of the OER is given in Chapter 8. Of course all calculations are based on optimum avalanche gain of the photodiode, which is in the range of M = 5 to 10.

As previously mentioned, the receiving filter either produces Gaussian or Nyquist pulses with "cosine roll-off" spectrum (roll-off factor, r = 1). Since both the transmitting pulse and the impulse response of the fiber are Gaussian-shaped, the receiving filter for producing Gaussian impulses is therefore also Gaussian. In order to suppress intersymbol interference produced by the pulse postcursors, a low-pass QFE can be connected in the case of Gaussian receiving filter. A bit error rate of $10^{-9}$ is assumed for all calculations. Calculations refer to ideal timing.

## Table 1
## THREE TYPES OF TRANSMISSION
## CHANNELS

| Type | $\lambda$ ($\mu$m) | $\alpha_F$ (dB/km) | $\tau_c$ (psec/nm $\cdot$ km) |
|------|------|------|------|
| 1 | 1.3 | 0.7 | 6 |
| 2 | 1.3 | 0.35[a] | $0 \leqslant \tau_c \leqslant 5$ |
| 3 | 1.55 | 0.2[a] | 15 |

[a] Ideal values (Chapter 3).

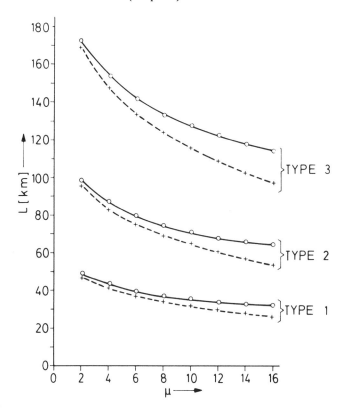

FIGURE 2. Optimum repeater spacing L for 1 Gb/sec systems as a function of the number of signal levels $\mu$ of the transmission codes used. (+) Receiver filter with Gaussian impulse response; (O) receiver filter for Nyquist equalization.

## B. Results

Table 1 shows the calculations carried out for three types of transmission channels. System 1 operates on a 1.3 $\mu$m wave-length range; the loss $\alpha_F$ and chromatic dispersion $\tau_c$ apply for one fiber which we used in our experiments (Chapter 10). Systems 2 and 3 correspond to ideal 1.3 or 1.55 $\mu$m systems.

Figure 2 shows the maximum attainable repeater spacing L for the three types of systems, as a function of the number of signal levels $\mu$ of a unipolar transmission code, with a transmission capacity of 1 Gb/sec. The relative transmission pulse duration is $\tau/T = 0.6$ ($\tau$ = impulse width, T = sampling interval). Calculations were carried out for Gaussian and Nyquist pulses at the input to the clocked comparator. From the calculations (Figure 2) we deduce:

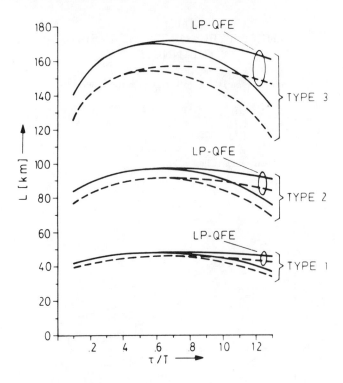

FIGURE 3.   Optimum repeater spacing L as a function of the relative transmission pulse width $\tau/T$ for binary 1 Gb/sec systems (—) and 2 Gb/sec systems ($\mu$). Receiver filter with Gaussian impulse response, with and without low-pass QFE.

- With attenuation-limited systems the maximum repeater spacings are achievable with binary transmission signals ($\mu = 2$).
- Roughly the same repeater spacings can be achieved for binary transmission signals with Gaussian and Nyquist receiving pulses. Since cascaded BB amplifiers closely approximate the Gaussian-shaped low pass, the additional expense for Nyquist receiving filters is not justified. Since Gaussian and Nyquist receiving pulses exhibit roughly the same horizontal eye pattern, both methods offer the same range of tolerance towards jitter in the sampling pulse during clock regeneration of the signals.

Thus, binary transmission signals and Gaussian receiving filters were made a precondition for all subsequent calculations. For transmission systems in the subscriber range where the question of the maximum attainable repeater spacing is less important, multistage transmission signals may offer certain advantages, because the reduced transmission speed imposes less demands on the electronic components required for signal processing purposes.

Figure 3 shows the maximum repeater spacing L for binary transmission signals and Gaussian receiver filters, as a function of the relative transmission pulse duration $\tau/T$.

The calculations for the three types of systems, were made in accordance with Table 1 in the 1 and 2 Gb/sec range. From Figure 3 we obtain the following deductions for attenuation-limited systems:

- Maximum repeater spacings are achieved for RZ transmission signals ($\tau/T = 0.5$ to 0.8).

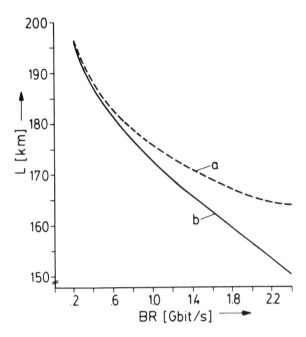

FIGURE 4. Optimum repeater spacing L for binary transmission with Gaussian receiving filter as a function of the bit rate (BR), using the chromatic dispersion as a parameter. Curve: (a) $\tau_c = 0$ psec/(nm·km); (b) $\tau_c = 20$ psec/(nm·km) and fiber-induced attenuation amounting to 0.2 dB/km (type ³).

- No improvement in repeater spacing can be achieved by means of low-pass QFE for RZ transmission signals. If other considerations (modulation behavior of the laser, required amplifier bandwidth) make NRZ transmission pulses necessary, the reductions in repeater spacing thus caused can be virtually avoided by the use of a low-pass QFE.

Figure 4 shows maximum repeater spacings for binary RZ transmission signals and Gaussian receiving filters as a function of the bit rate and of the chromatic dispersion of the fiber. Fiber loss is 0.2 dB/km. Curve a in Figure 4 applies for a dispersion-shifted fiber, operated at its zero dispersion point ($\tau_c = 0$; $\lambda = 1.55$ $\mu$m). Curve b applies for a normal, nonshifted fiber, operated at $\lambda = 1.55$ $\mu$m with a maximum chromatic dispersion of $\tau_c = 20$ psec/(km·nm).

With regard to 1.55 $\mu$m systems operated with DSM lasers, from Figure 4 we deduce:

- Dispersion-shifted fibers are not necessary. The increase in repeater spacing resulting from elimination of the chromatic dispersion only amounts to a few percent even in the case of 2 Gb/sec systems.

This statement applies only to DSM lasers. The restrictions on transmission capacity resulting from existing chromatic dispersion in the fiber for laser, with insufficiently suppressed sidelines in the spectrum (LMPN) are described in Chapter 10.

# REFERENCES

1. Personick, S. D., Receiver design for digital fiber optic communication systems, *Bell System Tech. J.*, 52, 843, 1973.
2. Muoi, T. V. and Hullett, J. L., Receiver design for multilevel digital optical fiber systems, *IEEE Trans. Commun.*, 23, 987, 1975.
3. Personick, S. D., Comparison of equalizing and nonequalizing repeaters for optical fiber systems, *Bell System Tech. J.*, 55, 957, 1976.
4. Smith, D. R. and Garrett, I., A simplified approach to digital optical receiver design, *Optical Quantum Electron.*, 10, 211, 1978.
5. Smith, R. G. and Personick, S. D., Receiver design for optical fiber communication systems, in *Semiconductor Devices*, 2nd Ed., Kressel, H., Ed., Springer-Verlag, Berlin, 1982, chap. 4.
6. Baack, C., Grenzrepeaterabstand digitaler optischer Übertragungssysteme hoher Bitrate, *Nachrichtentech. Z.*, 30, 65, 1977.
7. Baack, C., Optimierung eines optischen digitalen Übertragungssystems für 1,12 Gbit/s, *Frequenz*, 32, 16, 1978.
8. Krick, W. and Baack, C., Vergleich von "Nyquist" und "Partial Response" Übertragungsverfahren für digitale Lichtwellenleitersysteme, *Arch. Elektron. Übertragungstech.*, 35, 255, 1981.
9. Clark, A. P., *Principles of Digital Data Transmission*, Pentech Press London, Plymouth, 1976.
10. Baack, C., Grenzrepeaterabstand digitaler, optischer Übertragungssysteme bei unterschiedlichen Laserimpulsen, *Nachrichtentech. Z.*, 30, 577, 1977.

Chapter 3

# SINGLE-MODE FIBERS

Gerhard Elze, Heribert Münch, and Gerhard Wenke

## TABLE OF CONTENTS

I. Introduction....................................................................................................12

II. Properties of Fiber Materials ........................................................................12
    A. Material Attenuation..............................................................................12
    B. Material Dispersion ...............................................................................12

III. SM Fiber Waveguides......................................................................................14
    A. Dimensioning .........................................................................................14
    B. Field Distribution ..................................................................................14

IV. Transmission Properties..................................................................................15
    A. Chromatic Dispersion............................................................................15
    B. Birefringence..........................................................................................17
    C. Nonlinear Effects ..................................................................................17

V. Real Fibers .......................................................................................................17

References .................................................................................................................20

# I. INTRODUCTION

As already mentioned in Chapter 2, fibers employed for long-haul transmission systems of large transmission capacity can only be of single-mode (SM) type. Multimode (MM) fibers are not suitable for this purpose on account of bandwidth limitations due to mode dispersion and disturbances due to intermodal interference (modal noise).[1,2]

The authors do not intend to go into the basic principles of light propagation in dielectric waveguides and the exact theory of SM fibers within the scope of this book. Other books can be consulted for this purpose.[3-6]

Since the SM light waveguide constitutes the crux of the whole technology of optical broadband (BB) transmission, brief mention will at least be made of a few of the basic properties. In this connection, for the sake of brevity and clarity, simple approximations will be used.

# II. PROPERTIES OF FIBER MATERIALS

## A. Material Attenuation

Very pure quartz ($SiO_2$) serves as the starting material for present-day high-quality light waveguides. To vary the refractive index, the material is doped with Ge and P (to increase the refractive index) and B and F (to reduce the refractive index).

Optical losses in the material arise due to scattering and absorption of the light. Impurities involved in particular transition elements such as Cu, Fe, Ni, and Cr as well as OH-ions lead to high losses in the relevant wavelength range from 0.8 to 1.6 $\mu$m. Absorption losses, which as a fundamental principle cannot be avoided, occur in the UV and IR ranges due to absorption by the quartz material itself.

Scattering losses arise mainly at density fluctuations in the material whose periodicity is smaller than the wavelength of the light, i.e., so-called Rayleigh scattering. In general, scattering losses increase with increased doping of the material, among other factors. A further cause of scattering losses involves inhomogeneities in drawn waveguides which lead to radiation losses.

Figure 1 shows the causes of losses mentioned above as a function of the wavelength, using the example of a Ge-doped SM fiber.[7] In addition, the calculated resultant total loss is shown. The loss exhibits a minimum at about 1.55 $\mu$m.

## B. Material Dispersion

From the technological point of view in connection with transmission, a knowledge of the speed of propagation on the transmission path as a function of the wavelength is required.

For plane waves in an infinite stretch of material, the propagation constant $\beta$ is

$$\beta = k_0 \cdot n$$

where $k_0 = \dfrac{2\pi}{\lambda_0}$ ,    n = refractive index of the material, and
$\lambda_0$ = vacuum wave length.

To calculate the group delay time in relation to the length it is convenient first of all to define the group index $n_g$. The following is then obtained:

$$\tau = \frac{t_g}{L} = \frac{n_g}{c} = \frac{d\beta}{d\omega} \qquad \text{where } n_g = n - \lambda \frac{dn}{d\lambda} = n + \omega \frac{dn}{d\omega}$$

Figure 2 shows the curves of the refractive index n and the group refractive index $n_g$ in the wavelength range from 0.5 to 2 $\mu$m.[6]

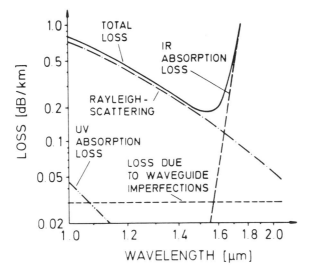

FIGURE 1. Representation of loss components in a Ge-doped SM fiber as a function of the wavelength.[7]

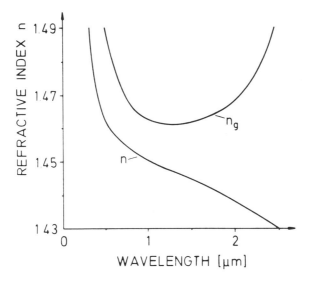

FIGURE 2. Refractive index n and group index $n_g$ of silica as a function of the wavelength.[6]

The dependence of the refractive index and thus the group delay time upon the wavelength is termed the material dispersion ($\tau_g$ which can be described approximately by the following equation:

$$\tau_g = \frac{d\tau}{d\lambda} = \frac{1}{c}\frac{dn_g}{d\lambda}$$

In the case of materials used for production of fibers, this function exhibits a zero-state near 1.3 μm, which is termed the zero-point of the material dispersion. In fact however, the terms of higher order neglected in the approximation produce an insignificant contribution to dispersion for practical purposes, particularly at this wavelength.

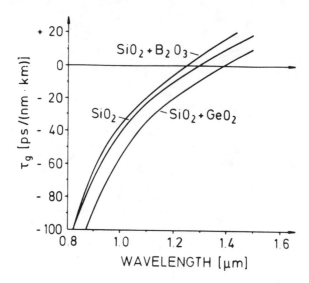

FIGURE 3.    Material dispersion of pure and doped silica vs. wavelength.[8]

Figure 3 shows the dispersion of pure and doped quartz as a function of the wavelength.[8] In optical fiber technology the dispersion is quoted as a change in delay time in picoseconds per nanometer wavelength change of the source and kilometer fiber length. As shown in Figure 3, the zero-point of the material dispersion can be shifted to higher wavelengths by doping the quartz with Ge and to lower wavelengths by doping with B.

## III. SM FIBER WAVEGUIDES

### A. Dimensioning

Figure 4 shows the scheme of construction of a light waveguide. The core with the refractive index $n_c$ and the cladding with the refractive index $n_{cL}$ are produced from correspondingly doped quartz material. To obtain a SM waveguide, specific dimensions must be adhered to for the core diameter 2a and the refractive index step ($n_c - n_{cL}$).

For this purpose the normalized frequency parameter

$$V = k_0 \cdot a \cdot NA$$

$$\text{with } NA = \sqrt{n_C^2 - n_{CL}^2}$$

is incorporated, where NA is the numerical aperture of the fiber for beam optics. For low doping percentages (low Rayleigh scattering) $n_c \cong n_{cL}$ so that:

$$NA = n_c\sqrt{2\Delta} \qquad \text{where:}$$

$$\Delta = \frac{n_C^2 - n_{CL}^2}{2n_C^2} \cong \frac{n_C - n_{CL}}{n_C}$$

### B. Field Distribution

The field distribution in a SM fiber is described in the core by the Bessel function $J_0$ and in the region of cladding by a field which decays exponentially. At the cut-off of the $HE_{21}$-mode (V = 2.4, see Section IV.A) approximately 20% of the power is carried in the cladding region. At lower values of V, correspondingly more power is carried in

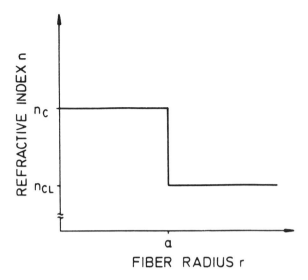

FIGURE 4. Schematic representation of a step-index optical waveguide.

the cladding. For this reason the cladding, in as far as it transmits light, must also be of a low-loss material so that fibers with overall low losses can be produced.

For practical applicatons, e.g., estimations of coupling losses, it is convenient to approximate the field by a Gaussian distribution:

$$E(r) = E(0) \exp\left(-\frac{r^2}{w_0^2}\right)$$

In this equation, E(0) is the field strength at the center of the core and w is the spot-size at which the field has decayed to the value E(0)/e. The error arising by describing the situation as a Gaussian distribution is insignificant for the majority of practical purposes.

## IV. TRANSMISSION PROPERTIES

For the practical employment of SM fibers in BB transmission systems, initial interest is mainly directed at the losses and dispersion. The loss properties of real fibers are treated in Section V.

The chromatic dispersion arises as a sum of the above-described material dispersion and the waveguide dispersion. In addition, under certain circumstances, birefringence effects must be taken into account.

The maximum transmittable power in the SM fiber is determined by nonlinear optical effects.

### A. Chromatic Dispersion

The optical field of the modes propagatable in a dielectric waveguide may extend with its tails into the range of the cladding material. The actual propagation constant of the mode thus lies between that of the core material and that of the cladding material so that the effective refractive index can be defined as:

$$n_{eff} = \frac{\beta}{k_0}$$

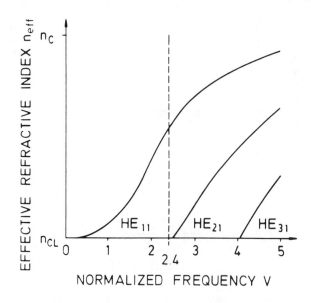

FIGURE 5.   Effective refractive index plotted against normalized frequency V.

Figure 5 shows the effective refractive index $n_{eff}$ as a function of the normalized frequency parameter V. The figure shows that the fundamental mode $HE_{11}$ can always propagate and the next higher mode begins to propagate at V = 2.405.

This provides a rule for the design of a SM waveguide, i.e., that at the application wavelength, V must be < 2.4; the SM fiber must be operated beneath the cut-off wavelength of the $HE_{21}$-mode.

If a given light waveguide is operated in the SM range and the wavelength of the light is changed slightly, V will change together with the propagation constant

$$\beta = k_0 \cdot n_{eff}(\lambda)$$

This dependence of the effective refractive index and thus the group delay time in the waveguide is termed the waveguide dispersion.

On calculation of the waveguide dispersion, it must be borne in mind that both the core and cladding materials exhibit material dispersion. Due to the differences in the doping percentages, the dispersion curves are different so that the refractive index difference is also a function of the wavelength as is in turn the normalized frequency V.

The chromatic dispersion $\tau_C$ of a SM fiber is given by the sum of the material and waveguide dispersion. The dependence of the group delay time referred to the length on the wavelength can be approximated for cases of weak guiding ($\beta_{CL} \cong \beta_C$) as follows:

$$\tau_C = \frac{d\tau}{d\lambda} = \frac{1}{C} \frac{dn_{CLG}}{d\lambda} - \frac{n_{CG} - n_{CLG}}{C\lambda} V \frac{d^2(VB)}{dV^2} \tag{1}$$

Here $n_{CLG}$ and $n_{CG}$ are the group indices of the core and cladding and

$$B = \frac{\beta - k_0 \cdot n_{CL}}{k_0 n_C - k_0 n_{CL}}$$

The first term in Equation 1 describes the already-known material dispersion; the second term describes the waveguide dispersion. By a judicious choice of the materials and the dimensions of the SM waveguide, it is possible to achieve compensation for the two terms for one or several wavelengths. Thus, zero-points in the chromatic dispersion occur. Furthermore, due to these compensation effects, it is possible to achieve low chromatic dispersion in the relevant spectral range of 1.3 to 1.6 $\mu$m as will be demonstrated later with the aid of an example.

## B. Birefringence

In real SM fibers, the circular symmetry in an ideal waveguide is upset by geometrical effects, e.g., by ellipticity of the fiber core and by mechanical strain. As a result, two orthogonal modes are no longer propagated with the same phase speed; their degeneration is destroyed. The difference in the effective refractive indices constitutes the birefringence. This causes the two orthogonal modes to have different propagation constants, thus producing dispersion.[9] Since the two modes cannot interfere, modal noise effects as are encountered in MM fibers do not arise.

With long fibers, the disturbing effects causing the birefringence are distributed irregularly over the fiber length so that the light transmitted changes its polarization randomly. The state of polarization at the end of the fiber is thus not predictable and is moreover often unstable, i.e., it changes with the temperature of the fiber, due to varying mechanical strain and with the wavelength of the transmitting source.[1] The dependence of birefringence on the wavelength in interplay with spectral changes in the semiconductor laser (see Chapter 4) leads to the disturbing effect that various fractions of an emitted light pulse suffer different group delay times. In contrast to pure pulse broadening due to dispersion, polarization distortions and noise occur as a result.[1,10,11] Additional noise arises if the transmission path contains parts with polarization-dependent attenuation.

For special applications in which no polarization distortions can be tolerated on the transmission path, e.g., in sensor technology or optical heterodyne systems, special polarization-maintaining SM fibers are thus employed[12,13] which however for baseband transmission technology as considered here, have hitherto not played an important role.

## C. Nonlinear Effects

A further problem in SM fiber transmission systems may arise as a result of nonlinear optical effects since the power densities in the fiber cores are very high and in the case of low-loss fibers large interaction lengths are possible. For optical sources of high coherence, stimulated Brillouin scattering[14] may occur in practice even for launched power in the milliwatt range. With sources of less coherence, Raman scattering occurs in the range of several hundred milliwatts and can lead to cross talk between channels in transmission systems with wavelength multiplex systems.[15] Thus in SM fiber systems the maximum transmittable power is limited due to nonlinear optical effects.

## IV. REAL FIBERS

In the practical design of SM fibers, numerous parameters have to be optimized which are not all independent of each other.[16,17] First of all, the cut-off wavelength must be selected to be shorter than the planned operational wavelength. This gives the normalized frequency parameter V and thus a set of possible refractive index differences and core diameters. If small refractive index differences and large core radii are selected, weakly guiding fibers are obtained which exhibit low attenuation but high microbending losses.[16] On account of the large spot-size, small splice losses also arise.

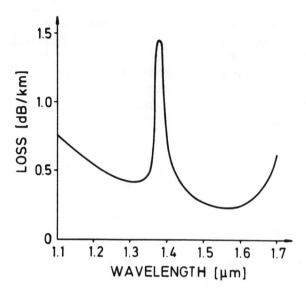

FIGURE 6.  Measured attenuation of a SM fiber plotted against wavelength.

On account of the low doping factor, the zero dispersion wavelength in addition lies close to that of quartz, i.e., at 1300 nm.

If on the other hand a fiber design is selected with a very large refractive index difference and a small core diameter, a strongly guiding fiber is obtained with low microbending sensitivity, higher attenuation (large Rayleigh scattering), and smaller spot-size which leads to higher average splicing losses.

In addition, because of the larger waveguide dispersion, the zero dispersion wavelength is shifted to longer wavelengths, in certain circumstances as far as the 1550 nm window.

In practice, structures are thus selected which lie between the extremes described. If approximately 0.5% is selected as a refractive index difference, a SM fiber is obtained which is easy to cable since it is not very sensitive to microbending. The attenuation values achievable lie in a range from 0.4 to 0.5 dB/km ($\lambda$ = 1300 nm), or 0.2 to 0.3 dB/km ($\lambda$ = 1500 nm).[16,18,19]

Figure 6 shows the measured attenuation curve of a SM fiber with "windows" in the 1.3 and 1.5 $\mu$m ranges. In between there is a range of high attenuation due to a very small amount of OH-ion contamination. The attenuation values in the two windows lie only slightly above the material values shown in Figure 1.

As a rule, the zero dispersion wavelength lies in the 1.3 $\mu$m window. In the 1.5 $\mu$m window, the chromatic dispersion then amounts to 15 to 20 psec/(nm · km).[16]

Particularly for long-haul transmission systems with large bandwidth × length products, utilization of the window with the minimum attenuation at 1500 nm is very interesting. With normal SM fibers however, problems arise due to the above-mentioned chromatic dispersion involving laser mode partition noise (LMPN) (Chapter 9) and pulse interference (Chapter 2) if no dynamic SM lasers are employed.

As already mentioned, it is possible to shift the zero point of the chromatic dispersion which occurs as a result of compensation of waveguide and material dispersion, by choice of the fiber design.

Figure 7 shows the material dispersion curve for silica, the waveguide dispersion of a step index fiber for two core diameters, and the related curves for the chromatic dispersion.[20] The figure shows how, in the case of step index fibers, the zero-point of the chromatic dispersion can be set between about 1.3 and 1.6 $\mu$m.

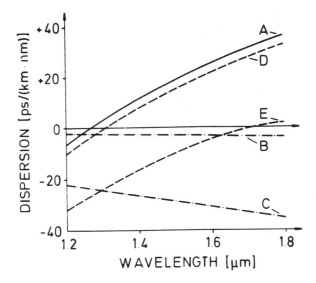

FIGURE 7.  Dispersion curves of step-index SM fibers.[20] (A) Material dispersion of fused silica; (B,C) waveguide dispersion of fibers with core diameters of 2a = 11 μm and 2a = 3.5 μm; (D,E) chromatic dispersion of fibers with core diameters of 2a = 11 μm and 2a = 3.5 μm.

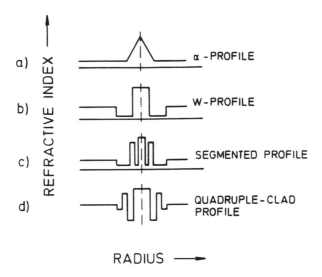

FIGURE 8.  Refractive index profiles of special SM fibers.

To overcome the problems arising with dispersion-shifted step index fibers involving higher attenuation and splice attenuation, attempts have been made to find other solutions. The most successful results were obtained hitherto in this connection from SM fibers with gradient index profiles exhibiting a triangular index profile (Figure 8a).

The spot-size at 1500 nm with these fibers is about the same size as with step index fibers in the 1300 nm range, so that similar coupling conditions prevail. Fibers of this type with a zero dispersion wavelength in the 1.5 μm window have been produced with an attenuation of below 0.25 dB/km at this wavelength.[16,21] Reproducible production of this fiber type with an attenuation below 0.3 dB/km has been reported.[22]

By even more complicated refractive index profiles of the fiber core (e.g., Figures 8b, 8c, 8d), several zero-points of the chromatic dispersion can be produced in the wavelength range from 1.3 to 1.6 $\mu m$. It is thus possible to keep the chromatic dispersion very small over the whole wavelength range, a feature which is important in particular for transmission systems with wavelength multiplex provisions. As an example, a quadruple-clad fiber — QC-fiber — (Figure 8d) has been produced which exhibits less than 2 psec/(nm·km) dispersion between 1.3 and 1.6 $\mu m$.[21] Production of a QC-fiber with an attenuation of 0.27 dB/km at 1.55 $\mu m$ has been reported.[23]

# REFERENCES

1. Epworth, R. E., Modal noise — causes and cures, *Laser Focus,* 17, 109, 1981.
2. Baack, C. et al., Modal noise and optical feedback in high-speed optical systems at 0.85 $\mu m$, *Electron. Lett.,* 16, 592, 1980.
3. Marcuse, D., *Light Transmission Optics,* (Bell Laboratory Series), Van Nostrand Reinhold, New York, 1972.
4. Unger, H. G., *Planar Optical Waveguides and Fibres,* Calrendon Press, Oxford, 1977.
5. Midwinter, J. E., *Optical Fibers for Transmission,* John Wiley & Sons, New York, 1979.
6. Grau, G., *Optische Nachrichtentechnik,* Springer-Verlag, Basel, 1981.
7. Miya, T. et al., Ultimate low-loss single-mode fibre at 1.55 $\mu m$, *Electron. Lett.,* 15, 106, 1979.
8. Fleming, J. W., Material dispersion in lightguide glasses, *Electron. Lett.,* 14, 326, 1978.
9. Kapron, F. P., Borelli, N. F., and Keck, D. B., Birefringence in dielectric optical waveguides, *IEEE J. Quant. Electron.,* QE-8, 222, 1972.
10. Petermann, K., Nonlinear transmission behaviour of a single-mode fiber transmission line due to polarisation coupling, *J. Optical Commun.,* 2, 59, 1981.
11. Enning, B. and Wenke, G., Demonstration of coloured noise and signal distortions in baseband spectra of broadband optical transmission systems, *NTZ-Arch.,* 5, 301, 1983.
12. Rashleigh, S. C. and Stolen, R. H., Preservation of polarization in single-mode fibers, *Laser Focus,* 19, 155, 1983.
13. Okamoto, K., Varnham, M. P., and Payne, D. N., Polarization-maintaining optical fibers with low dispersion over a wide spectral range, *Appl. Opt.,* 22, 2370, 1983.
14. Cotter, D., Stimulated Brillouin scattering in monomode optical fiber, *J. Optical Commun.,* 4, 10, 1983.
15. Chraplyry, A. R. and Henry, P. S., Performance degradation due to stimulated Raman scattering in wavelength-division-multiplexed optical-fibre systems, *Electron. Lett.,* 19, 641, 1983.
16. Midwinter, J. E., Monomode fibres for long haul transmission systems, *Br. Telecom. Technol. J.,* 1, 5, 1983.
17. Kitazama, K.-I., Kato, Y., Ohashi, M., Ishida, Y., and Uchida, N., Design considerations for the structural optimization of single-mode fiber, *J. Lightwave Technol.,* LT-1, 363, 1983.
18. Bachmann, P., Leers, D., Lennar, M., and Wehr, H., Preparation of single-mode fibers by the low pressure PCVD process, *Proc. ECOC 83 — 9th Eur. Conf. Optical Commun.,* Melchior, H. and Sollberger, A., Eds., Elsevier, Amsterdam, 1983, 5.
19. Miyamoto, M., Akiyama, M., Shiota, T., Sanada, K., and Fukuda, O., Fabrication and transmission characteristics of VAD fluorine doped single-mode fibers, *Proc. ECOC 83 — 9th Eur. Conf. Optical Commun.,* Melchior, H. and Sollberger, A., Eds., Elsevier, Amsterdam, 1983, 9.
20. White, K. I. and Nelson, B. P., Zero total dispersion in step-index monomode fibres at 1.3 and 1.5 $\mu m$, *Electron. Lett.,* 15, 396, 1979.
21. Cohen, L. G., Mammel, W. L., Jang, S. J., and Pearson, A. D., High-bandwidth single-mode fibers, *Proc. ECOC 83 — 9th Eur. Conf. Optical Commun.,* Melchior, H. and Sollberger, A., Eds., Elsevier, Amsterdam, 1983, post-deadline paper.
22. Anislie, B. J., Beales, K. J., Cooper, D. M., Day, C. R., and Nelson, B. P., The reproducible fabrication in long length of dispersion shifted single-mode fibres with ultra low loss, *Proc. ECOC 83 — 9th Eur. Conf. Optical Commun.,* Melchior, H. and Sollberger, A., Eds., Elsevier, Amsterdam, 1983, 53.
23. Bachmann, P., Geittner, P., Herman, W., Lydtui, H., Rau, H., Ungelenk, J., and Wehr, H., Recent progress in the preparation of GI- and SM-fibres by means of the PCVD process, *Technical Digest* 29A5-3, in 4th Int. Conf. Integrated Opt. Optical Fiber Commun., Tokyo, 1983.

Chapter 4

# OPTICAL SOURCES FOR BROADBAND (BB) TRANSMISSION SYSTEMS

Gerd H. Grosskopf, Peter Meissner, Erwin Patzak, Godehard Walf,
and Gerhard Wenke

## TABLE OF CONTENTS

I.      Introduction.................................................................22
        A.      Requirements for Optical Source Devices.....................22
        B.      Fundamental Properties of Semiconductor Lasers............22

II.     Dynamic Properties .......................................................25
        A.      Modulation Behavior ..............................................25
        B.      Spectral Behavior in the Case of Modulation ...............29
        C.      Laser Modulation Circuits........................................32
                1.      Input Impedance of the Laser .........................32
                2.      Modulation Circuits......................................33

III.    Intensity Noise.............................................................35

IV.     Optical Feedback .........................................................39
        A.      Influence of Reflections within the Compound Cavity Mode
                Region..........................................................40
        B.      Influence of Reflections within the Light Injection Mode
                Region..........................................................43

V.      Dynamic Single-Mode Laser (DSM Laser).........................43
        A.      DFB and DBR Lasers..............................................44
        B.      Coupled Cavity Lasers............................................46
        C.      Stabilization by Means of an External Mirror ...............46
        D.      Injection Locking...................................................47

References .........................................................................50

# I. INTRODUCTION

## A. Requirements for Optical Source Devices

Semiconductor lasers are employed as transmitters in optical broadband (BB) systems (bit rate 400 Mb/sec), because their compactness and their high level of efficiency make them superior to other light sources, and moreover, in comparison with light emitting diodes (LEDs), they can be modulated at higher bit rates, have a narrower emission spectrum, and their radiation can be coupled more efficiently into fibers.

There are already numerous books and articles[1-5] dealing with semiconductor lasers and consequently we shall restrict ourselves here to the properties that are important for BB transmission systems with direct intensity modulation — which forms the subject of this book. More detailed exposition of individual problems can be found in the bibliography, but it should be borne in mind that the list is by no means complete.

The requirements established for application of lasers in a BB transmission system can be summarized as follows:

1. The lasers must be capable of modulation up to sufficiently high frequencies (several gigahertz) with high efficiency.
2. The overall power consumption should be low (low threshold, high efficiency).
3. The intensity noise should be as low as possible.
4. The lasers should exhibit single-mode (SM) characteristics even in the case of modulation (not essential, if the transmission wavelength corresponds to the zero dispersion wavelengths of the fibers [see Chapter 10]).
5. There should be little sensitivity to optical feedback in the laser.
6. The far field pattern should allow a high coupling efficiency into the monomode fiber.

## B. Fundamental Properties of Semiconductor Lasers

The semiconductor III-V compounds $Ga_{1-x}Al_xAs$ and $In_{1-x}Ga_xAs_yP_{1-y}$ are used to manufacture semiconductor lasers. The emission wavelength of $Ga_{1-x}Al_xAs$ lasers is in the region of 0.8 $\mu$m. The material system $In_{1-x}Ga_xAs_yP_{1-y}$ has emission wavelengths of between 0.92 and 1.65 $\mu$m depending on the composition. This is a particularly suitable wavelength range because the zero dispersion length of the optical silicon fiber used occurs at about 1.3 $\mu$m, and its minimum loss occurs at about 1.5 $\mu$m.

The materials are grown epitactically on GaAs or on InP substrate, respectively. In this way multilayer structures can be produced, each layer of which has a different composition. The transition between two such layers, each of which can in addition be doped differently, is referred to as a heterojunction. In the case of $In_{1-x}Ga_xAs_yP_{1-y}$ it is necessary for the lattice constants of the layers to correspond, and this is an additional condition for the composition of the layers. If the lattice constants do not match, tensions and dislocations can occur in the layer structure, which have an adverse effect on the lifetime of the laser and on its ability to function properly. This requirement is not of critical importance in the case of $Ga_{1-x}Al_xAs$, since the lattice constant is approximately independent of Al content x, because Al and Ga have a virtually identical[1] atomic radius.

The possibility for the manufacture of such heterostructures forms the basis for the fact that the current densities required for laser operation can be reduced to levels that permit continuous operation.[1,3] In this way it is possible to set up the laser-active (optically amplifying) layer between n- and p-type confinement layers with a greater energy gap. Heterobarriers are formed in direction x (see Figure 1), which prevent the outflow of the charge carriers injected by current I into the active region. Consequently these charge carriers must recombine within the layer itself. In addition, an optical

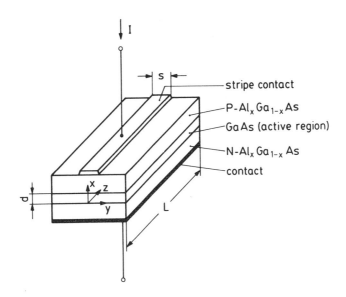

FIGURE 1.    Basic structure of a semiconductor laser.

layer waveguide is formed as a result of the confinement layers. Thus, the distribution of intensity of the confined lightwave and of the gain are adjusted to one another, producing high effective gain of the confined lightwave. Normally the geometrical length L of the laser amounts of values between 200 and 400 $\mu$m; the thickness d of the amplifying (active) layer is approximately 0.1 $\mu$m (see Figure 1). The laser reflectors are normally formed by the cleaved facettes of the chips (Fabry-Perot resonator).

The semiconductor lasers can be classified according to the way in which the transversal confinement of the lightwave within the active region (y direction in Figure 1) is undertaken. Lasers without such a confinement tend towards instability, e.g., in the form of filament formation and the initiation of higher transversal modes, leading to so-called "kinks", i.e., pronounced nonlinearities in the relationship between light output and input current. The resulting need for a transversal SM characteristic and the need for a further reduction in losses, and thus for a drop in the current densities, mean that the lightwave and the carriers must be confined in direction y. This is particularly important in the case of $In_{1-x}Ga_xAs_yP_{1-y}$ lasers, because this material exhibits a pronounced nonradiating recombination, which can be attributed among other reasons to the Auger effect and intervalence band absorption.[6] This nonradiating recombination is highly dependent upon temperature and increases along with the charge carrier density at a greater rate than the gain. For this reason it is necessary to keep the losses in the laser as low as possible, in order to manage with the least possible gain.

The simplest possibility for confinement is to design the electrical contact as a narrow stripe with a width of several micrometers. The charge carriers are then only injected into the active region over a narrow range.[7,8] In this way a gain profile is produced in the active layer in which the lightwave is amplified inside the center region and absorbed outside. This means of guiding the lightwave is referred to as gain guiding. Two structures of this kind are shown in Figure 2. Lasers of this type have a high threshold current of 80 to 300 mA, due to high absorption losses in the external range.

An additional step is implemented, e.g., in the case of channeled substrate planar (CSP) lasers, the structure of which is shown in Figure 3a. Here too, confinement of the charge carrier takes place in the manner described above; however, a stripe wave guide is produced with the aid of effective refractive index jumps from the layer wave guide in y direction. This refractive index jump is used to guide the lightwave. This

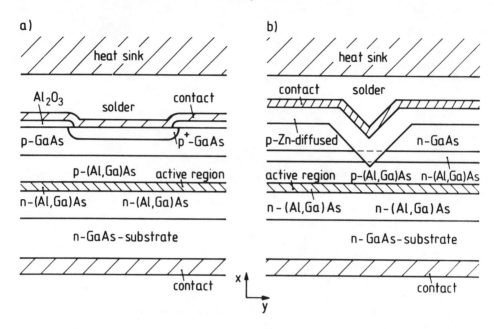

FIGURE 2.    Structure of gain-guided lasers. (a) Stripe geometry laser; (b) V-groove laser.

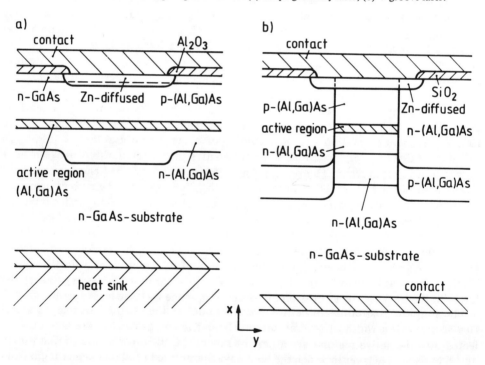

FIGURE 3.    Structure of index-guided laser. (a) CSP laser; (b) BH laser.

type of guidance is referred to as index guiding. The buried heterostructure laser (BH laser) (Figure 3b)[9] forms a further step. Here the active region is surrounded by material with a lower refractive index. The charge carriers are also guided transversally by means of potential barriers. Due to the ideal concentration of the current inside the active region, combined with effective index guidance for the light wave, this structure permits a very low threshold current (10 to 20 mA) to be achieved. However, against these advantages must be weighted the technological difficulties involved in the manufacture of this complicated structure.

In addition to the laser structures described, other possibilities exist for achieving favorable wave and charge carrier guiding, e.g., transverse junction stripe lasers.[10]

## II. DYNAMIC PROPERTIES

### A. Modulation Behavior

The semiconductor lasers available at the present time can be modulated up to several gigahertz. Using special types of lasers, cut-off frequencies of more than 10 GHz are achieved.[11,12] The modulation behavior is of major importance for use in optical communication links. A great many articles have been published dealing with this subject.[13-33]

The dynamic response of the optical output power to changes of the impressed current depends on the dynamic behavior of the laser process itself as well as on the electrical properties of the laser diode, such as the lead inductance and capacity of the laser package. The dynamic behavior of the laser process is described in terms of rate equations for the charge carrier concentration in the active region N(y,t) and the intensity $S_j(t)$ of the light in the individual modes. These rate equations are classical approximations of quantum mechanical equations. They have the following form.[15]

$$\frac{\partial N(y,t)}{\partial t} = \frac{J(y,t)}{ed} - \frac{N(y,t)}{\tau_{sp}} + D \frac{\partial^2 N(y,t)}{\partial y^2}$$

$$- \sum_j G_j(N[y,t]) |E_j(y,t)|^2 S_j(t) \tag{1}$$

$$\frac{dS_j(t)}{dt} = \left\{ \int G_j(N[y,t]) |E_j(y,t)|^2 \, dy - \frac{1}{\tau_{pj}} \right\} S_j(t)$$

$$+ \frac{\beta_j}{\tau_{sp}} \int N(y,t) |E_j(y,t)|^2 \, dy \tag{2}$$

Because the active region is usually very thin (d = 0.1 $\mu$m) only one local variable y is required, i.e., the one that is parallel to the layer structure. This thickness is only slight compared with the diffusion length (5 $\mu$m) of the charge carriers, and for this reason no variation in charge carrier density can be expected in a vertical direction to the layer structure. Since it may be assumed that the lasers are transversally SM (in direction y in Figure 1) as mentioned above, the index j distinguishes only between various longitudinal modes. J(y,t) is the current density injected into the active region, and e is the elementary charge. D (= 40 cm²/sec) is the diffusion constant and $\tau_{sp}$ (= 1 nsec) is the lifetime of the charge carriers due to spontaneous emission and nonradiating recombination processes. $G_j(N[y,t])$ is the effective gain for mode j with the normalized shape $E_j$ (y,t). $\tau_{pj}$ (= 2 psec) is the lifetime of the photons in the resonator. It is obtained from the reflection factor of the end surfaces and from the length of the lasers. $\beta_j$ (= $10^{-3}$ to $10^{-5}$) indicates the degree of spontaneous emission coupled into the mode j. All numerical values given in brackets are typical for GaAs lasers.

Analysis of the rate equations forms the basis for an understanding of the laser dynamic behavior, provided one is not interested in the phase of the light and in the linewidth. This is true despite the fact that the equations in the form given above already represent a simplified model. Thus, for example, transport of the charge carriers cannot generally be described my means of a diffusion equation with diffusion

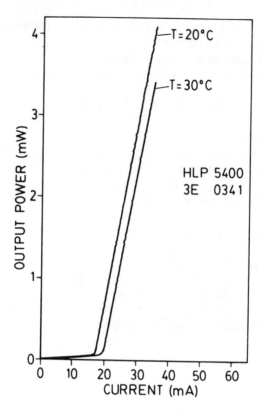

FIGURE 4.    Power current characteristic for a BH
laser operating at a wavelength of 1.3 μm at temperatures of 20 and 30°C.

constant independent of the carrier density. The dependence of the recombination
processes on the density of the charge carriers is a more complicated matter.

In most cases the rate equations 1 and 2 are discussed in an even simpler form, i.e.,
as equations for the charge carrier number in the active region and the photon number
in the resonator, with the result that effects produced by the spatial distribution of the
charge carrier density and mode intensity are ignored. At least as far as the dynamic
behavior of index-guided lasers is concerned, a simplified model such as this provides
a sufficiently accurate qualitative description.

The rate equations can also be used in a discussion of stationary behavior. This is
described in terms of an output power-input current characteristic. Figure 4 shows
measured characteristics at 20 and 30°C. The difference between these two curves indicates the temperature-dependency of the material properties. The two most important parameters of the characteristic are its slope and the threshold current $I_{th}$. Above
the threshold current, which is between 10 mA for BH lasers and up to 300 mA for
gain-guided lasers, the output power virtually increases in a linear way. The typical
slope is between 0.1 and 0.2 mW/mA. Up to 0.3 mW/mA[34] has been achieved. The
carrier density, and consequently the gain below the threshold, increases almost line-
arly with increasing current. However, for steady-state operation the gain cannot exceed the resonator losses. Thus, above threshold the carrier density remains nearly
constant (clamped) at the value at which the gain compensates the resonator losses.

One way of obtaining an insight into the dynamic behavior of the laser is to investigate its small signal behavior. To do this a small modulation current is superimposed
on the constant bias current. Figures 5 and 6 can be achieved theoretically with the aid

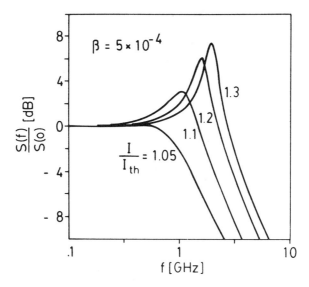

FIGURE 5. Frequency response for a semiconductor laser for various operating points.

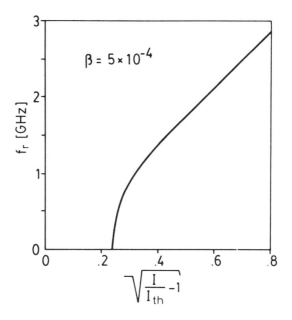

FIGURE 6. Resonance frequency of the frequency response of a semiconductor laser as a function of the operating point.

of a simple SM rate equation model. Figure 5 shows the frequency response of the intensity for various operating points above the threshold current, where $I/I_{th}$ is the ratio between bias current and threshold current. One noticeable feature of the curves is the resonance peak at just a few gigahertz. The frequency of this resonance is displayed as a function of the bias current in Figure 6. It increases proportional to $(I_{th}-1)^{1/2}$ for higher values of I. Due to the highly simplified nature of the rate equation model, both illustrations should only be interpreted as examples of qualitative behavior.

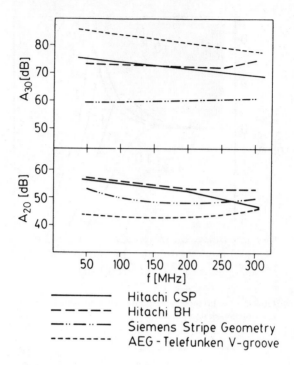

FIGURE 7.     Nonlinearities of second and third order
for various laser diodes.[25]

Besides the operating point, there exist additional factors determining the height of
the resonance peak. Two major factors consist of diffusion of the charge carriers and
the rate of spontaneous emission in the laser mode. The stronger the influence of diffusion, and the greater the coupling of spontaneous emission in the laser modes,[16,17]
the weaker the resonance. Multimode (MM) lasers generally show a weaker resonance
than SM lasers, because the number of spontaneously emitted photons increases with
the number of modes contributing to the overall intensity.[13]

Electrical equivalent circuit diagrams have been developed for small signal behavior,[18,19] providing a way of including external circuitry of the laser in the analysis (see
Section II.C.1).

Small signal behavior only provides incomplete information about the dynamic behavior of the laser. In the case of the pulse levels and bit rates used in digital BB
transmission systems, the semiconductor laser cannot be described in terms of a linear
system. For this reason it is necessary to investigate the large signal behavior.

An investigation of nonlinearity using sinusoidal modulation signals provides valuable information. In this context a variety of theoretical[20-22] and experimental[23-25] work
has been carried out. In such investigations measurements were made of harmonics of
the second and third order, and of the intermodulation products. Figure 7[25] shows
nonlinear distortions of the second and third order for a number of different types of
lasers. Here the lasers were investigated by the well-known two-tone method, with a
modulation signal at a fixed frequency of 7 MHz while the other varied between 50
and 300 MHz. It is shown[23] that the nonlinearities are sufficiently low at frequencies
under about 500 MHz. This is also the range in which the dynamic behavior can be
described by the linear static characteristic.[23] With increasing frequency up to the resonance frequency, the response amplitude of the carrier density inside the active region
increases. This produces a nonlinear behavior, a fact that is immediately apparent in
the rate equations.

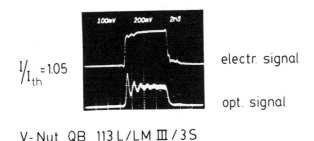

$I/I_{th} = 1.05$    electr. signal

opt. signal

V- Nut QB 113 L / LM Ⅲ / 3 S

FIGURE 8.    Pulse response of a V-groove laser showing high
relaxation oscillations.

Another possibility is to investigate the large signal behavior in the case of digital modulation. In this case the pulses are superimposed on a constant bias current. Figure 8 displays typical electrical input and optical output pulse shapes. The optical pulse shape is characterized by a sharp rise, followed by relaxation oscillations, which can be extremely pronounced according to the type of laser. If the bias current (in this case the lower level of the pulses) is well below the threshold, the rise is delayed by a few nanoseconds, because the first part of the current pulse is required to produce sufficient carrier density, i.e., sufficient optical gain in the active region for the stimulated emission.

### B. Spectral Behavior in the Case of Modulation

The spectrum of a modulated semiconductor laser may differ significantly from that of an unmodulated laser.[26-28] A distinction can be drawn between different types of spectral changes.

1.    The MM character of the spectrum becomes more pronounced.
2.    Mode jumps occur.
3.    The spectral lines become frequency-modulated.

For a time-averaged spectrum, behaviors 1 and 2 show up in a broadening of the spectrum. The distinction between 1 and 2 only becomes apparent in a time-resolved spectrum. Due to fiber dispersion, the first two points restrict the bandwidth that can be achieved.

This is of importance in particular for transmission systems using wavelengths of 1.5 $\mu$m, because even though there is minimal loss at such wavelength (0.2 dB/km), dispersion amounts to 16 psec/(km·nm). Given a dynamic spectral width of about 5 nm, a bandwidth length product of only 10 Gb/sec·km could be attained.[28]

Section V deals with the possibilities for achieving SM characteristics even in directly modulated operation. Without employing special frequency-selective measures, mode selection occurs solely on the basis of the frequency dependency of the optical gain. This may be sufficient for high side mode suppression in the case of CSP lasers in the short wavelength region (at $\lambda = 0.85$ $\mu$m).

Spectral changes are caused by the changes in temperature of the active region, and by the changes of carrier density in the active region occurring with modulation.

The increase in temperature produces a reduction in the energy gap of the semiconductor material. This results in a shift of the gain maximum to higher wavelengths.[35] In the case of GaAs the temperature-dependence amounts to about 2.5 Å/K. An increase in the carrier density causes the gain profile to shift, due to the intensified filling up of the conduction and valence bands, respectively, thereby selecting modes with

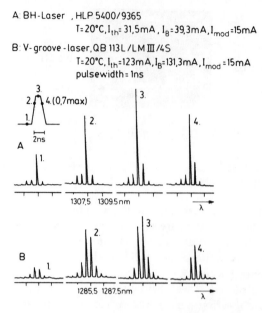

A: BH-Laser , HLP 5400/9365

T=20°C, $I_{th}$= 31,5mA , $I_B$=39,3mA , $I_{mod}$=15mA

B: V-groove-laser, QB 113L /LM III/4S

T=20°C, $I_{th}$=123mA, $I_B$=131,3mA, $I_{mod}$=15mA

pulsewidth= 1ns

FIGURE 9.    Time-resolved dynamic spectra at different points of a pulse with 1 nsec halfwidth.[26] (A) BH laser; (B) V-groove laser.

shorter wavelengths.[35] In the case of modulation, the carrier density and the temperature inside the active region changes, shifting the wavelength of maximum gain. At modulation frequencies below 10 MHz the changes in temperature are quite significant, while carrier density changes are relevant at higher frequencies. This variation in the maximum gain wavelength can lead to mode hopping.

Figures 9 and 10 are intended to demonstrate the behavior described above.[26] Figure 9 shows the time-resolved spectra for a BH laser and a V-groove laser. The modulation signal was a 1.12 Gb/sec NRZ pseudorandom bit sequence with a word length of $2^{15} - 1$ b. The spectra were recorded at the marked points of a single pulse taken from the bit pattern. Only slight spectral changes can be observed both for the BH laser as well as for the V-groove laser. This is due to the fact that in the example shown, the lower signal level is only slightly above the threshold, and there is only a small change in the carrier density between the different points of the pulse. With a current below the threshold we could expect pronounced changes in the spectrum, since the carrier density will then change appreciably between the clamped value and the value belonging to the working point below threshold. Intensified changes in the spectrum occur, e.g., if a longer sequence of "1"s occur in the signal, because in this case the active region warms up and consequently mode hopping to larger wavelengths occurs. This phenomenon is shown in Figure 10, which displays the spectrum at different points of a pulse with a length of 7 nsec. It is apparent that the temperature changes are significant when the bias current is above threshold, but with the high bit rates observed here this is only effective in the case of long sequences of "1" or "0". In contrast, with bias currents below threshold, the transient changes in charge carrier density lead to a widening of the spectrum.[29] For these reasons RZ modulation and bias current above threshold are advantageous regarding the spectral behavior.[29] However, one must be prepared for a less favorable extinction ratio.

Additionally, frequency modulation of individual longitudinal modes is superimposed on the intensity modulation. This frequency modulation is based on the dependence of the refractive index of the laser material on temperature and charge carrier

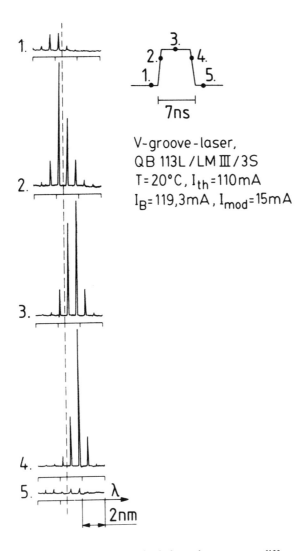

FIGURE 10.    Time-resolved dynamic spectra at different points of a pulse with 7 nsec halfwidth for a V-groove laser.[26]

density. An increase in temperature produces increase in the refractive index, while an increase in the charge carrier density produces a decline in the refractive index. This leads to smaller or larger frequencies, respectively, for the longitudinal mode. As a result of this frequency modulation the spectral width of the individual line may be significantly larger than the spectral width of the modulation signal.[30,31]

Figure 11 shows the frequency modulation spectrum for individual types of lasers.[30] This behavior can theoretically be obtained from the frequency response of the charge carrier density, and from the temperature-dependence of the refractive index. The resonance peak corresponds to the peak in the intensity frequency response, which was dealt with in the previous section. An explanation for the rise of low frequencies can be found in connection with temperature-induced effects, and the flattened-out curve in the case of medium frequencies is attributed to the infuence of diffusion within the active layer.[30]

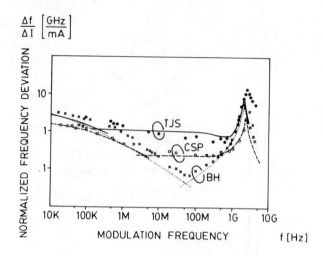

FIGURE 11.    Normalized frequency deviation for TJS, CSP, and BH lasers vs. modulation frequency.[30]

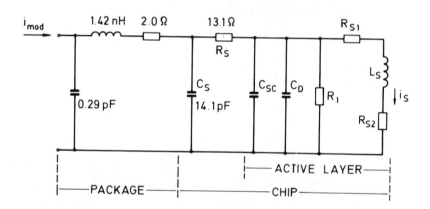

| I (mA) | $C_{SC}$ (pF) | $C_D$ (pF) | $R_1$ ($\Omega$) | $R_{S1}$ (m$\Omega$) | $R_{S2}$ ($\mu\Omega$) | $L_S$ (pH) |
|--------|--------------|-----------|-----------------|---------------------|-----------------------|-----------|
| 20 | 10 | 380 | 1.23 | 23.4 | 34.0 | 7.07 |
| 25 | 10 | 381 | 0.829 | 24.1 | 11.8 | 4.19 |
| 30 | 10 | 382 | 0.628 | 24.7 | 6.0 | 3.01 |

FIGURE 12.    Small signal equivalent circuit of a packaged BH laser HLP 3400.[32]

## C. Laser Modulation Circuits

### 1. Input Impedance of the Laser

Determination of the input impedance and of the modulation behavior of the laser, and the deduction of an equivalent electrical circuit diagram[18,19,32,33] are of particular interest in the development and simulation of laser modulation circuits.

Figure 12 shows the small signal equivalent circuit diagram for a semiconductor laser (BH, Hitachi® HLP 3400)[32] together with the parasitic elements of the package. The

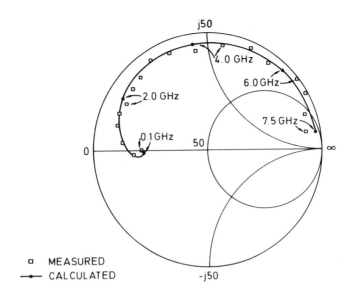

FIGURE 13.    Measured and calculated input impedance of the BH
laser biased above threshold.[19]

equivalent circuit diagram for the small signal behavior of the active layer of the laser
chip is derived from the rate equations of the laser. In this model the current i, in the
series circuit $R_{s1}$, $L_s$, and $R_{s2}$ corresponds to the light output power. The values of the
components describing the active layer depend on the laser current I. They are given in
the table with Figure 12 for various currents above the threshold current for a BH
laser.

The capacitance between the contacts of the laser chip and the bulk series resistance
are modeled by means of $C_s$ and $R_s$. The equivalent circuit diagram for the laser package includes the capacitance of the connecting pin and the inductance and resistance
of the bond wire.

With a laser current above the threshold current the impedance of the active layer
compared with the impedance to the remaining network (parasitic elements of the chip
and package) is extremely low. Thus, to determine the input impedance the active layer
can be replaced by a short circuit. In contrast to the components of the equivalent
circuit of the active layer, the values of the parasitic chip and package elements are
independent of the laser current, and thus also independent of the level of modulation. They are determined with the aid of $S_{11}$ — parameter measurement ($I > I_{th}$) —
and a network analysis program (computer-aided fitting).[19] The Smith diagram (Figure
13) shows the measured and calculated input impedance vs. the frequency for the BH
laser.[19]

## 2. Modulation Circuits

This section describes modulation circuits suitable for high-rate modulation of a
laser diode. Because of the direct relation between current and light power in the laser
and the low laser impedance with an inductive part, the modulation stage should be a
source with a high internal impedance in comparison with the input impedance of the
laser.

The current switch shown in Figure 14 is well suited for digital modulation. The
transistors T1 and T2 are switched alternately from the on to the off state, depending
on the data signal. In this way no bias current flows through the modulation circuit.
There is thus a distinct separation of the modulation current $i_{mod}$ and bias current $I_B$.

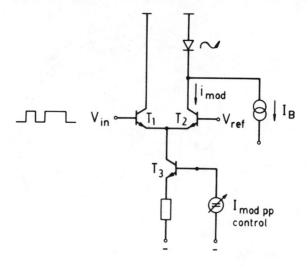

FIGURE 14.    Current switch as modulation circuit.

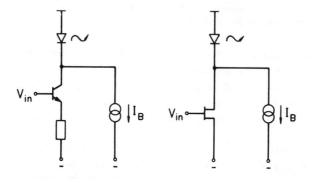

FIGURE 15.    Modulation circuit with a bipolar transistor or field effect transistor.

The amplitude $I_{modpp}$ of the modulation current can be controlled via the current source (transistor T3).

One advantage of this type of circuit is the facilities it offers for integration, and the prospect for common integration of the modulation stage and laser on one substrate.

The two modulation circuits shown in Figure 15, with a bipolar transistor and a field effect transistor, respectively, are suitable both for digital and analogue modulation, provided they are operating in the linear range.

Within digital transmission systems they can be used, e.g., in cases where preequalization of the modulation signal is carried out, and thus a quasi-analogue modulation of the laser is required. A disadvantage inherent in both circuits is that part of the laser bias current flows via the transistor, and therefore it is affected by the transistor (e.g., temperature drift). Moreover, compared to the current switch, the modulation current cannot be controlled independently from the bias current.

Due to the possible spatial separation of signal source and laser, the circuit shown in Figure 16 is suitable for experimental set-up in the laboratory. The modulation signal is fed to the laser with a 50 $\Omega$ coaxial line. Matching impedances of the laser and the cable is achieved by means of a series resistor $R_s$ (chip resistor). Because of the imaginary component of the input impedance of the laser, optimum BB matching is not achieved. To avoid multiple reflections, the signal source should have an output imped-

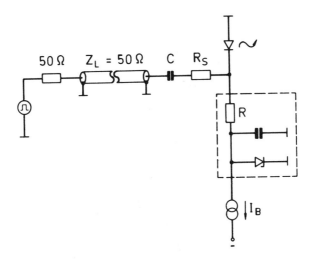

FIGURE 16.    Laser modulation using a coaxial line.

ance of 50 Ω. The source impedance, seen from the laser, amounting to 50 Ω + R, is thus relatively high in relation to the low input impedance of the laser. By means of the capacitor C (chip) connected into the signal path, the DC components of the signal source and bias current source are separated. For laboratory operations this circuit arrangement offers the following advantages:

1.    The laser modulation circuit only requires a few passive components.
2.    Digital as well as analogue modulation are possible. There are no restrictions regarding the choice of modulation signal (pulse form, amplitude, etc.).
3.    Due to the AC coupling of signal source and laser, the signal source can be easily changed.

The network in the dashed lines protects the laser from voltage or current spikes generated by the bias current source.

In the case of AC coupling it should be remembered that a baseline shift of the modulation signal occurs, provided the power density spectrum of the modulation signal contains a DC component. This baseline shift corresponds to a change in the bias current or operating point. To keep this effect as low as possible, the distribution of "low" and "high" states in the data signal should be kept as even as possible and long blocks of "low" and "high" states should be avoided. Furthermore, problems arise in determining the actual bias current $I_B$. It can only be obtained by taking into account the peak-to-peak value $I_{modpp}$ of the modulation current, the signal shift behind the coupling capacitor, and the current $I_B$ (Figure 17).

## III. INTENSITY NOISE

There is a large volume of literature devoted to the subject of intensity fluctuations of semiconductor lasers.[36-41] Fluctuations occur in the number of charge carriers and in the number of photons as a result of stimulated and spontaneous emission processes, and of absorption processes and radiation losses. By approximation these fluctuations can be described for a large number of charge carriers and photons with the aid of rate equations in combination with noise driving forces:[36,42]

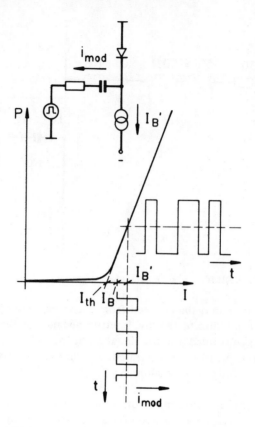

FIGURE 17.    Basic circuitry and operating point
adjustment for a laser modulation above the threshold with AC coupling of modulation stage and laser.

$$\frac{d\hat{N}(t)}{dt} = \frac{I(t)}{e} - \frac{\hat{N}(t)}{\tau_{sp}} - \sum_j G_j(\hat{N}[t]) \, S_j(t) + F_N(t) \qquad (3)$$

$$\frac{dS_j(t)}{dt} = \left\{ G_j(\hat{N}[t]) - \frac{1}{\tau_{pj}} \right\} S_j(t) + \frac{\beta_j}{\tau_{sp}} \, \hat{N}(t) + F_{sj}(t) \qquad (4)$$

Equations 3 and 4 correspond to those given in Section II (Equations 1 and 2), the
symbols marked with a ∧ representing the number of carriers or photons in the active
layer. The noise driving forces $F_N(t)$ and $F_{sj}(t)$ contained in these equations have shot
noise character, and are used to describe the influence of generation rate and recombination rate fluctuations of the carriers, or alternatively, the influence of emission
rate and absorption rate fluctuations of the photons. With the aid of Equations 3 and
4 it is possible to obtain the noise output spectrum of the unmodulated laser by means
of linearization around the operating point, and by calculating the variation in intensity of the individual modes.[36]

Figure 18 indicates the calculated and measured relative intensity noise (RIN) density
of all modes in the case of a GaAlAs-CSP laser, as a function of the operating point.[36]
Because Equations 3 and 4 correspond to Equations 1 and 2 used to describe the modulation behavior, we also obtain similar resonance curves to those with which we are
familiar from Section II.

When investigating the intrinsic noise of the semiconductor laser, it should also be
remembered that fluctuations in very low intensity side modes also make a considerable

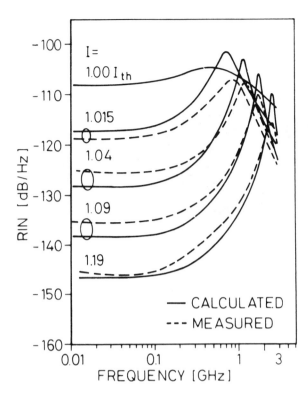

FIGURE 18.    Measured and calculated RIN spectrum of a
CSP laser for different working points.[36]

contribution to the total noise output. While investigating the noise intensity of individual modes it is also apparent that these are anticorrelated below the resonance frequency of the noise spectrum. This is because all modes draw their energy from one single reservoir, the amplifying medium. This leads to competition between the individual modes, which exchange energy with each other via the charge carrier number. Figure 19 shows the spectrum of a GaAlAs stripe geometry laser and the correlation of the two main modes $U_1$ and $U_2$. To get this figure the fluctuations of the powers of these modes are shown as amplitudes of the x- and y-deflections of an oscilloscope, respectively. The brightness distribution on the screen is then an image of the joint probability distribution of the power fluctuations of the two modes. In this measurement the deflections of both axes were normalized, to achieve the same variance for both modes. Equal phase correlation would produce an ellipsoid with the longer principle axis at an angle at 45° and an antiphase correlation and ellipsoid with the principle axis at an angle of 135°.

As a result of the antiphased correlation, the noise components of all modes compensate for frequencies below that of the resonance frequency. Consequently, the signal-to-noise ratio (SNR), when determined in the total spectrum, is up to 40 dB greater than that of an individual spectral line. Figure 20 shows the RIN

$$\frac{\overline{|\Delta S(\omega)|^2}}{S^2}$$

spectrum of intensity of the dominant mode, and the overall intensity. It can be seen that considerable noise enhancement arises when only the dominant mode is detected, although in the case of this laser the side modes are relatively small.

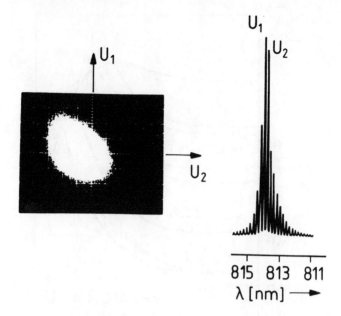

FIGURE 19.     Spectrum and correlation between the two main modes for a GaAlAs stripe geometry laser.

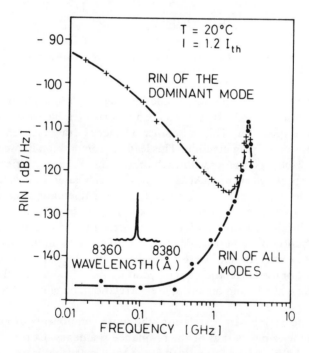

FIGURE 20.     RIN spectrum of the dominant mode and of all modes of a GaAlAs CSP laser.[36]

Due to the fiber dispersion, the modes pass the fiber at different velocities, so that the anticorrelation is more and more lost with increasing fiber length, leading to a strong increase in noise power. This produces a considerable deterioration in the SNR. Similar deterioration in the SNR also occurs when individual modes experience different transmission losses. Detailed investigations on this so-called laser mode partition noise (LMPN) can be found.[36,43-47]

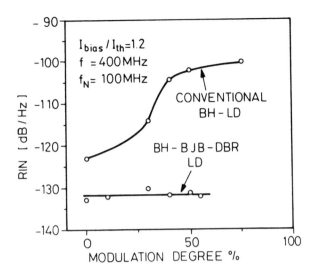

FIGURE 21.    RIN at 100 MHz ($f_N$) of main mode as a function of modulation depth for a conventional BH and DBR laser[48] (f = modulation frequency).

Figure 28 of Section V.A shows the spectrum of a conventional BH laser compared with the spectrum of distributed Bragg reflector (DBR) laser, for the modulated and unmodulated case. The RIN for the main mode of each type is given in Figure 21.[48] It can be seen that in the case of the BH laser, the main mode noise increases with the degree of modulation, while in the case of the DBR laser the RIN is independent of the degree of modulation.

The influence of digital modulation on the noise behavior of the laser was examined[43-44] using a simplified model, disregarding the dynamics of the laser. The correlation between two modes was described by the statement:

$$\overline{(S_j[t] - \overline{S_j[t]}) (S_i[t] - \overline{S_i[t]})} = - k^2 \, \overline{S_i(t)} \, \overline{S_j(t)} \tag{5}$$

and it was assumed that there is no fluctuation in the total output power per pulse. The factor $k^2$ serves to characterize the laser, and can be determined from measurements. The main advantage of this description is in the static modeling of the competition noise which permits simple approximate calculation of the distance-bandwidth products.[49] A model based on the rate equations is given[50] which is able to describe the increasing decorrelation between the modes for transmission over long distances via a monomode with a dispersion of 90 psec/(nm·km). In Figure 22 the RIN of all modes is given as a function of the distance for a GaAlAs V-groove laser. It can be seen that a considerable increase in noise occurs as a result of decorrelation due to dispersion of the fiber as the fiber length increases.

## IV. OPTICAL FEEDBACK

The properties of the semiconductor laser are influenced to a significant extent by optical feedback, i.e., by feedback of a fraction of the output power caused by reflection of the light at the ends of the fiber, at the connectors or other reflecting surfaces along the transmission line. Such feedback can be selectively applied in order to influence the properties of the laser as required. For example, it is possible to achieve a DSM state (see Section V.B) or to reduce the line width.[51-55] However, uncontrolled

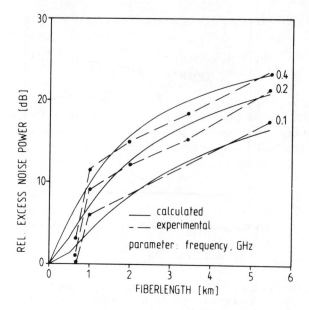

FIGURE 22.     Laser mode partition noise for a gain-guided laser against SM fiber length.

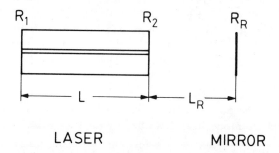

FIGURE 23.     Principal arrangement of a semiconductor laser with external reflector.

feedback can cause excessive intensity noise,[56-60] mode hopping,[61-63] or instabilities,[61-64] deteriorating the performance of the system. The principle arrangement of a semiconductor laser with optical feedback is shown in Figure 23.

An external mirror with reflection factor $R_R$ is installed at a distance $L_R$ from the laser facette. $R_1$ and $R_2$ are the reflection factors of the laser facettes. The significant parameters in this setup are the internal and external circulation time $\tau_{in}$ and $\tau_{ex}$, i.e., the times taken by the light to circulate once in the laser, or to pass along the space $L_R$ twice, respectively. A distinction is drawn between the various types of feedback, according to whether the external round trip time is long or short compared with the coherence time of the light. The range of values $\tau_{ex}$ for which $\tau_{ex}$ is smaller than the coherence time is called "compound cavity mode region". Such reflections occur in optical transmission systems, e.g., at the front end of the fiber, or at nearby connectors. The range of values for $\tau_{ex}$ larger than the coherence time is called "light injection mode region". This superimposition in optical transmission systems is caused, e.g., by reflections at more remotely located connectors or on the far end of the fiber.

## A. Influence of Reflections within the Compound Cavity Mode Region

Properties of the laser diodes such as the power current characteristic,[62-65] the spectrum[66,67] noise,[57-60,65] modulation behavior[65] are strongly influenced by the type of feedback. In this respect one significant parameter is the round trip time in the external resonator in relation to the round trip time within the laser. If the round trip time in the external resonator is long compared with the round trip time in the laser, we refer to a long resonator, and if it is shorter it is a short resonator.

The round trip time $\tau_{in}$ in the laser cavity and the round trip time $\tau_{ex}$ in the cavity formed by the external mirror and the laser facette adjacent to it as well as the reflectivities $R_1$, $R_2$, and $R_R$ determine the structure of the spectrum of this coupled cavity.

The light propagated back into the laser at $R_2$ is the sum of the light directly reflected at this mirror and of light coming back after one or several round trips in the external resonator. This effect of the external mirror can be described by a complex facette reflectivity, which is a periodic function on frequency with period $1/\tau_{ex}$.

The spectrum of the semiconductor laser can then be calculated with the aid of the MM rate equations, using the frequency-dependent reflection factor, which leads to frequency-dependent phase changes of the light reflected at the facette.

$$-2\pi j + 2\pi f_j \frac{2L \cdot n(N,f_j)}{c} = \text{Im}\left(\frac{1}{2} \ln [R_1 \, r(f_j)^2]\right)$$

$$r(f) = \frac{\sqrt{R_2} + \sqrt{R_R}\, e^{-j2\pi f \tau_{ex}}}{1 + \sqrt{R_2 R_R}\, e^{-j2\pi f \tau_{ex}}}$$

(6)

The phase condition for constructive superposition is given in Equation 6. In the case of weak feedback this formula simplifies considerably and the feedback can be described by only one parameter, namely the so-called feedback parameter, as for instance derived in Reference 70.

$$X = \sqrt{\frac{R_R}{R_2}}\,(1 - R_2)\,\frac{\tau_{ex}}{\tau_{in}}$$

In Figure 24a the spectrum of a laser without feedback is shown. A side mode suppression of about 10 dB can be identified, which is solely due to the frequency-dependence of the gain. In Figure 24b the resonator losses of a laser with an external mirror at a distance corresponding to $\tau_{ex} = \tau_{in}/2$ and an effective reflectivity of 4% are depicted. Additionally, the gain is shown and the mode frequencies (solutions of Equation 6) are marked. For the central (lasing) mode the resonator losses and the gain nearly compensate each other. The mode suppression for the side modes is given by the difference between resonator losses and gain at the corresponding frequencies. In this example, their difference and thus the suppression is especially high for the modes directly neighboring the central mode, whereas modes with more remote frequencies may experience relatively little suppression. Figure 24c depicts the spectrum of the laser with the external mirror described by the resonator losses shown in Figure 24b.

Section V.C discusses the best choice of spacing and reflection factor. The arrangement of the short resonator as described above provides effective suppression of the side modes. For this reason it is used in order to stabilize MM lasers, and thus to maintain the influence of the dispersion at a particularly low level in the case of optical BB transmission (see Chapter 10).

If the reflection point is so far away from the laser that the external round trip time is greater than the internal round trip time, i.e., if a long resonator is present, the mode distribution that occurs is certain to differ. The spectrum shown in Figure 25 is that of a laser with an external resonator, with the reflector located approximately 10 cm from the laser, and with a 4% degree of reflection. Altogether each mode is separated into a bundle of satellite modes, grouped around the original mode with a spacing of ap-

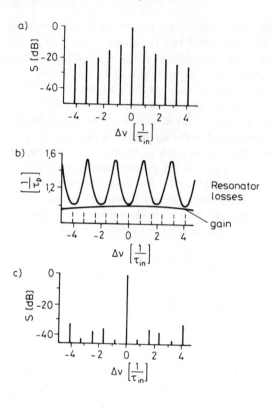

FIGURE 24.    (a) Spectrum of solitary laser; (b) gain
and resonator losses for a laser with a short external
resonator; (c) the spectrum of a laser with a short external resonator. $\Delta v$ is measured from the main mode frequency.

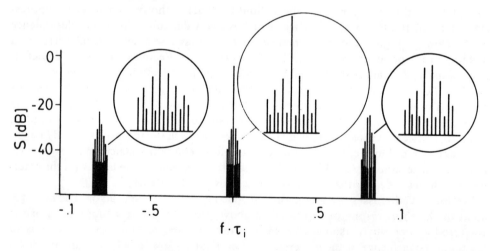

FIGURE 25.    Spectrum of a laser with a long external resonator.

proximately $1/\tau_{ex}$. Reflections which manifest themselves in this way occur in optical
transmission systems, e.g., in the case of connectors on the transmitting side.

Since in the case of external reflectors the losses are strongly frequency-dependent,
small changes of the internal round trip time due to refractive index and gain variations
can lead to strong changes of the laser properties. For example, since gain and refrac-

tive index are both temperature-dependent, changes in temperature can lead to mode hopping[61,62] and self-pulsations. Apart from temperature variations, phase changes in the reflected field may also cause mode hoppin or instability.[63]

The modulation and noise properties are also affected by feedback. On the one hand, the resonance frequency of the small signal frequency response shifts to low or high frequencies according to the phase condition existing between the reflecting field and light within the lasers on the other hand, resonances occur at those frequencies which correspond to a multiple of the external round trip time. The greater the feedback, the more pronounced these resonances become. The same feature appears in the intensity noise power spectrum.[65]

## B. Influence of Reflections within the Light Injection Mode Region

In the case of reflections occurring at distances which can be considered large compared with the coherence length, the reflected light can be interpreted as fluctuating signal injected from another coherent source. The behavior of the laser is not simply explained by the fact that light with a fluctuating intensity is injected into the laser, but additionally we must take into account the phase relationship (interference) between the transmitted and received light.[57,58] The phase of the field in the laser is synchronized with the injected field as in the case of injection locking (see also Section V.D). If frequency fluctuations in the frequency of the laser or changes in the phase of the reflected light occur, the synchronization can be lost. In the synchronized state the output power is larger than in the nonsynchronized state. The reaction of the laser to reflections within the injection mode region can be conceived of as a change between the synchronized and nonsynchronized state. Thus frequency fluctuations are converted into intensity fluctuations. This leads to additional noise of up to 20 dB in the light output.[58]

## V. DYNAMIC SINGLE-MODE LASER (DSM LASER)

As described in Section II, mode hopping and a multiplicity of modes occur when semiconductor lasers are modulated with high bit rates. As a result of fiber dispersion, both effects lead to a reduction in the maximum bit rate or in the field length. This applies in particular to 0.8 and 1.5 $\mu$m transmission links; however, it can also be a significant factor in a 1.3 $\mu$m system if the zero dispersion wavelength and the transmitted wavelength differ (see Chapter 10). It is therefore necessary to stabilize the laser at one single longitudinal mode in modulated operation in order to increase the capacity of a transmission system. Examples of how successful such measures can be have been shown in experimental transmissions over distances of 134.23 km with 445.8 Mb/sec using a distributed feedback (DFB) laser[71] over a distance of 161.5 km with 420 Mb/sec with a $C^3$ laser[72] and with 445.8 Mb/sec over 170 km with injection locking.[73]

There are various ways of achieving a DSM state, i.e., through frequency-selective feedback, coupled resonators, injection locking, and short laser cavities. The group of lasers with frequency-selective feedback include those lasers in which the reflection on the laser facettes is replaced partially or completely by reflection at integrated gratings (DFB/DBR laser). The group of lasers with coupled resonators include the grooved and cleaved coupled cavity lasers and the lasers with an external resonator formed by mirrors or gratings. In all configurations the secondary modes show higher losses than the main mode. It can be estimated that the difference between gain and loss must be at least 5 to 10 cm$^{-1}$ (see Reference 28) for the side modes, in order to reduce the side mode intensity to values less than 1% of the main mode intensity.

We shall not deal separately with lasers with a short active cavity. This feature is only used due to the fact that the longitudinal mode spacing in the case of short resonators is greater, which means that the frequency-dependency of the gain is sufficient for side mode suppression.

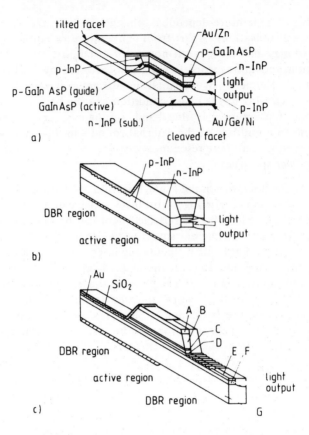

FIGURE 26.     (a) Principle structure of a DFB laser;[74] (b) principle structure of a DBR laser with split surface;[75] (c) principle structure of a DBR laser with two gratings;[75] A — p-GaInAsP cap layer; B — p-InP cladding layer; C — p-GaInAsP antimeltback layer; D — $Ga_x In_{1-x}As_yP_{1-y}$ active layer; E — n-InP separation layer; F — n-$Ga_uIn_{1-u}As_yP_{1-y}$ output guide; and G — n-InP substrate.

## A. DFB and DBR Lasers

Figure 26a to c shows the basic structure for DFB and DBR lasers.[74,75] The gratings are etched in corrugations in the surface of the active (DFB) or passive (DBR) waveguide. The Bragg wavelength of the gratings determines the emission wavelength of the laser. The effectiveness of the coupling achieved by the gratings is determined not only by the material parameters, but also by the shape and depth of the corrugations.

In the case of DFB lasers, the light can be coupled out by means of a cleaved facet (as shown in Figure 26a) or by a waveguide. This is important in achieving the integration of such a laser within an integrated optical circuit. The same applies to the DBR laser which can be manufactured with one (Figure 26b) or two gratings (Figure 26c). DFB and DBR lasers are implemented both in the shortwave (0.8 $\mu$m)[76,77] as well as in the longwave range length.[78,79] Whereas fully symmetrical DFB or DBR lasers basically have a spectral two-mode behavior, preference can be given to one mode by means of an asymmetrical arrangement of the gratings in the case of the DBR laser, or by using different reflector facettes in the case of the DFB laser.

The reason for the SM characteristic for such an asymmetric DFB laser is clarified by Figure 27.[74] This shows the gain required in order to achieve the laser threshold for individual DFB resonance modes (structure as in Figure 26) as a function of the devia-

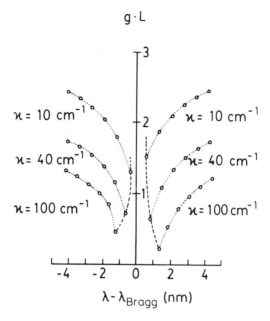

FIGURE 27. Required mode gain as a function of the difference between the laser and Bragg wavelength[74] for an asymmetric DFB laser. $\varkappa$ = coupling coefficient of the grating; L = length of the active region (structure as Figure 26a).

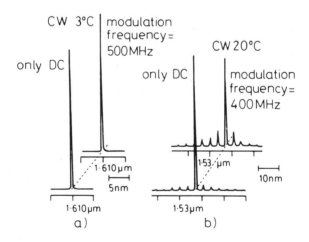

FIGURE 28. Comparison of the spectra between DBR laser (a) and BH laser (b) during modulation.[48] The intensity scale is linear.

tion of the wavelength from the Bragg wavelength. The parameter of the curves is the coupling coefficient $\varkappa$. This DFB laser shows a preference for the modes with longer wavelengths than the Bragg wavelength. This is achieved by means of different reflection factors at the two laser facettes. The losses of the side modes are so high that their intensity remains extremely low compared with that of the main mode.

Both DFB and DBR lasers demonstrate SM behavior even under modulation, up to frequencies of 1.9 GHz.[80] Figure 28[48] demonstrates the DSM character of a DBR laser.

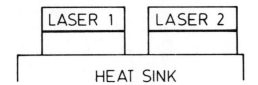

FIGURE 29.     Structure of a C³ laser.

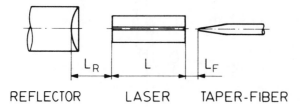

REFLECTOR          LASER          TAPER-FIBER

FIGURE 30.     Principle arrangement for a laser stabilized by an external reflector.

In contrast to the spectra of a conventional BH laser, the spectrum of the DBR laser does not display any recognizable increase in intensity in its side modes when modulated.

The construction of the grating and in the case of DBR lasers, the achievement of an effective coupling between the active region waveguide and the DBR waveguide, causes severe technological difficulties in comparison with the normal Fabry Perot type lasers.

### B. Coupled Cavity Lasers

Coupled cavity lasers consist of two coupled laser sections. The first of these sections acts as a light source, while the other serves as an active etalon or as a modulator. A distinction is drawn between groove coupled cavity lasers[81] and cleaved coupled cavity lasers (C³ lasers).[72] The principle arrangement of a C³ laser is shown in Figure 29. In the case of groove coupled cavity lasers a furrow is etched diagonally in a laser and the two sections provided with their own contacts. A C³ laser is obtained by cleaving a laser into two parts, both parts usually being unequal. The two parts are then lined up. Because both parts of the coupled cavity lasers can be controlled separately, this type of laser is notable for a large number of possible operating states. Temporally resolved spectral measurements reveal that coupled cavity lasers can be operated dynamically in a SM.[72] Compared with DFB and DBR lasers, one disadvantage of the coupled cavity laser is that the DSM characteristic is not inherent, but must be achieved by appropriate setting of the operating point.

### C. Stabilization by Means of an External Mirror

As already dealt with in Section IV, external reflectors can be used to influence the modulation behavior,[70,82,83] properties of the overall spectrum (DSM operation),[51,67,69,84] and in the case of longer resonators, the linewidth of the laser modes.[52-55] Various types of reflector are used, e.g., concave and flat mirrors[84,85] or gratings placed in a Littrow arrangement.[70,82,86-88] We shall only be concerned with arrangements that permit DSM operation. For this kind of operation, external resonators with short reflector spacing are suitable,[84,85] i.e., the reflector spacing being less than the optical length of the laser resonator. Figure 30 shows such an arrangement permitting DSM operation. Figure 31 shows the spectra of an InGaAsP/InP laser in

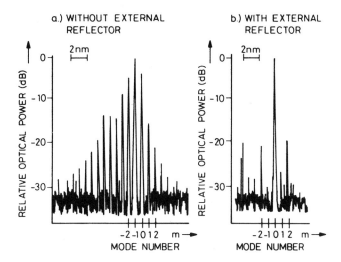

FIGURE 31. Spectrum of a 1.3 $\mu$m BH laser with and without an external reflector.

CW operation (a) without stabilization and (b) with a reflector. The spherical reflector is placed at a distance (approximately 600 $\mu$m) equivalent to about half the optical laser resonator length from the rear laser mirror. This ratio of the length ensures that each second mode is particularly well suppressed. The various side modes are 20 dB below the main mode. The spectra can be calculated with the aid of the model described in Section IV for various parameters of the reflector arrangement. Figure 32 shows the ratio of output power in the largest side mode and the output power in the main mode as a function of the reflector spacing. It is assumed that the mirror is adjusted so that the main mode has minimum loss. For a laser with an active region length of 300 $\mu$m as in regard ot mode suppression, optimum spacing of 200 $\mu$m is achieved, which corresponds to $\tau_{ex}/\tau_{in} = 0.19$.

The measured DSM behavior is shown in Figure 33. In this case the configuration is the same as that in Figure 30. In both types of modulation (1 Gb/sec RZ, 1 Gb/sec NRZ) no significant increase in the side mode intensity is to be seen.

Such a laser stabilized by a reflector is not inherently DSM; in fact the external reflector must be adjusted to fractions of the light wavelength, and if there are fluctuations or changes in the laser, e.g., as a result of aging, subsequent adjustment or control is necessary.

### D. Injection Locking

To achieve DSM behavior, light is injected from a reference laser (master) into a laser (slave). The mode of the slave laser with a frequency sufficiently close to the mode of the master laser is drawn towards the frequency of the master. In this state the slave laser oscillates synchronously with the master laser.[89-91] This locked mode is emphasized at the expense of the side modes, achieving DSM operation.[91] This procedure was originally used as a means of improving modulation properties (suppression of relaxation oscillations)[92,93] of laser diodes.

Figure 34 displays this principle of coherent light injection. The light from the master laser is coupled into the second laser. The optical isolator placed between the lasers prevents feedback to the master. If the master laser possesses a MM spectrum, the master and slave must possess different resonator lengths, e.g., different mode spacings, to ensure that the resonator frequencies of both coincide approximately at only one frequency. Another method is the selection of the required mode using an Etalon placed between the two lasers.

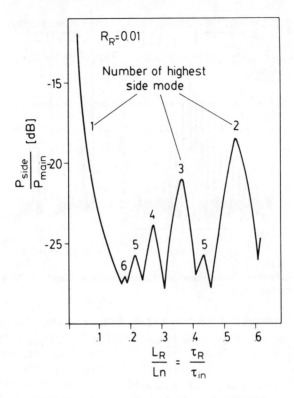

FIGURE 32. Output power of side mode with highest intensity as a function of the distance between laser and reflector. n = the refractive index of the active region.

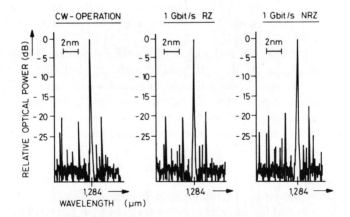

FIGURE 33. Spectra of a 1.3 μm BH laser stabilized with the aid of an external mirror for CW operation and under modulation.

The effect of injection locking is shown in Figure 35.[85] Both spectra, i.e., that of the master laser and that of the slave laser, are of MM character. The modes marked with a star are coupled with each other. This produces a spectrum of the locked slave, in which the ratio of the side mode power to the main mode power is less than −25 dB. The time average spectra of directly modulated slave lasers (1 Gb/sec RZ and 1 Gb/sec NRZ) are shown in Figure 36, for the locked and unlocked slave laser. The pronounced suppression of the side modes under coherent injection can be seen. Deterio-

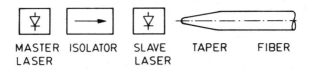

FIGURE 34.   Principal configuration for injection locking.

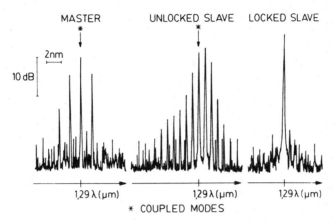

FIGURE 35.   Spectra of the master, unlocked, and locked slave laser.

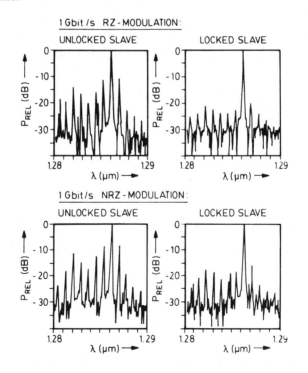

FIGURE 36.   Spectra of unlocked and injection locked laser at RZ and NRZ modulation.

ration in the mode suppression is greater under modulation than in the case of the laser stabilized with the aid of an external mirror.

One disadvantage of this arrangement is that the frequency of each laser must be adjusted in order to remain in the "locked" state. Moreover, two lasers and an isolator are required, which is a sophisticated solution.

# REFERENCES

1. Casey, H. C. and Panish, M. B., *Heterostructure Lasers,* Academic Press, New York, 1978.
2. Grau, G., *Optische Nachrichtentechnik,* Springer-Verlag, Heidelberg, 1981.
3. Kressel, H. and Butler, J. K., *Semiconductor Lasers and Heterojunction LED's,* Academic Press, New York, 1977.
4. Kressel, H., Ettenberg, M., Wittke, J. P., and Ladany, I., Laser diodes and LEDs for fiber optical communication, in *Semiconductor Devices* (Topics in Applied Physics), Kressel, H., Ed., Springer-Verlag, Heidelberg, 1982, 39.
5. Thompson, G. H. B., *Physics of Semiconductor Laser Devices,* John Wiley & Sons, New York, 1980.
6. Kamiya, T. and Kamata, N., Temperature characteristics of InGaAsP laser diodes and light emitting diodes. Optical devices and fibers, *Jpn. Annu. Rev. Electron. Comput. Telecommun.,* 5, 59, 1983.
7. Lengyel, G., Meissner, P., Patzak, E., and Zschauer, K. H., An analytic solution of the lateral current spreading and diffusion problem in narrow oxide stripe (GaAl)As/GaAs lasers, *IEEE J. Quantum Electron.,* 18, 618, 1982.
8. Joyce, W. B., Current crowding and carrier confinement in double heterostructure lasers, *J. Appl. Phys.,* 51, 2394, 1980.
9. Chinone, N., Saito, K., Ito, R., and Aiki, K., Highly efficient (GaAl)As buried-heterostructure lasers with buried optical guide, *Appl. Phys. Lett.,* 35, 513, 1979.
10. Hirano, R., Oomura, E., Ikeda, K., Matasai, K., Ishii, M., and Susaki, W., AlGaAs TJS lasers with very low threshold current and high efficiency, *Jpn. J. Appl. Phys.,* 17(1), 355, 1978.
11. Lau, K. Y., Bar-Chaim, N., Ury, I., and Yariv, A., A 11 GHz direct modulation bandwidth GaAlAs window laser on semiinsulating substrate operating at room temperature, OFC 84, post deadline paper.
12. Lin, C., Burrus, C. A., Eisenstein, G., Tucker, R. S., Besomi, P., and Nelson, R. J., 10 GHz Pico-second pulse generation in gain-switched short cavity InGaAsP injection lasers by high frequency direct modulation, OFC 84, post deadline paper.
13. Arnold, G., Russer, P., and Petermann, K., Modulation of laser diodes, in *Semiconductor Devices* (Topics in Applied Physics), Kressel, H., Ed., Springer-Verlag, Heidelberg, 1982, 39.
14. Adams, M. J., Rate equations and transient phenomena in semiconductor laser, *Opto-Electron.,* 5, 201, 1973.
15. Channin, D. J., Botez, D., Neil, C. C., Conolly, J. C., and Bechtle, D. W., Modulation character-istics of constricted double-heterojunction AlGaAs laser diodes, *IEEE J. Lightwave Technol.,* 1, 146, 1983.
16. Ito, M., Ito, T., and Kimura, T., Dynamic properties of semiconductor lasers, *J. Appl. Phys.,* 50, 6168, 1979.
17. Furuya, K., Suematsu, Y., and Hong, T., Reduction of reasonancelike peak in direct modulation due to carrier diffusion in injection lasers, *Appl. Opt.,* 17, 1949, 1978.
18. Harder, C., Katz, J., Margalit, S., and Yariv, A., Noise equivalent circuit of a semiconductor diode, *IEEE J. Quantum Electron.,* 18, 333, 1982.
19. Tucker, R. S. and Pope, D. J., Microwave circuit models of semiconductor injection lasers, *IEEE Trans. Microwave Theor. Tech.,* 31, 289, 1983.
20. Harth, W., Large-signal direct modulation of injection lasers, *Electron. Lett.,* 9, 532, 1973.
21. Harth, W., Properties of injection lasers at large-signal modulation, *AEÜ,* 29, 149, 1975.
22. Ikegami, T. and Suematsu, Y., Large signal characteristics of direct modulated semiconductor injection lasers, *Electron. Commun. Jpn.,* 53B, 69, 1970.
23. Stubkjaer, K. E., Semiconductor lasers, their linearity and noise properties, Report LD45, Electro-magnetics Institute, Technical University, Lyngby, Denmark, 1981.
24. Hong, T., Suematsu, Y., Chung, S., and Kang, M., Harmonic characteristics of laser diodes, *J. Opt. Commun.,* 3, 42, 1982.
25. Grosskopf, G. and Kueller, L., Measurement of nonlinear distortions in index- and gain-guiding GaAlAs lasers, *J. Opt. Commun.,* 1, 15, 1980.
26. Wenke, G. and Enning, B., Spectral behavior of InGaASP/InP 1.3 $\mu$m lasers and implications on the transmission performance of broadband Gbit/s signals, *J. Opt. Commun.,* 3, 122, 1982.
27. Petermann, K., Theoretical analysis of spectral modulation behaviour of semiconductor injection lasers, *Opt. Quantum. Electron.,* 10, 233, 1978.
28. Suematsu, Y., Long-wavelength optical fiber communication, *Proc. IEEE,* 71, 692, 1983.
29. Selway, R. and Goodwin, A. R., Effect of dc bias level on the spectrum of GaAs lasers operated with short pulses, *Electron. Lett.,* 12, 25, 1976.
30. Kobayashi, S., Yamamoto, Y., Ito, M., and Kimura, T., Direct frequency modulation in AlGaAs semiconductor lasers, *IEEE J. Quantum Electron.,* 18, 582, 1982.
31. Olesen, H. and Jacobsen, G., A theoretical and experimental analysis of modulated laser fields and power spectra, *IEEE J. Quantum Electron.,* 18, 2069, 1982.

32. Tucker, R. S. and Pope, D. J., Circuit modelling of the effect of diffusion on damping in a narrow-stripe semiconductor laser, *IEEE J. Quantum Electron.*, 19, 1179, 1983.

33. Figueroa, L., Slayman, Ch. W., and Yen, H.-W., High frequency characteristics of GaAlAs injection lasers, *IEEE J. Quantum Electron*, 18, 1718, 1983.

34. Burkhard, H. and Kuphal, E., InGaAsP mushroom stripe lasers with low cw threshold and high output power, *Jpn. J. Appl. Phys.*, 22, L721, 1983.

35. Stern, J., Calculated spectral dependence of gain in excited GaAs, *J. Appl. Phys.*, 47, 5382, 1976.

36. Jaeckel, H., Lichtemissionsrauschen und dynamische Verhalten von GaAlAs-Heterostruktur- Diodenlasern im Frequenzbereich von 10 MHz bis 8 GHz, Dissertation, University of Zurich, 1980.

37. Haug, H., Quantum mechanical rate equations for semiconductor lasers, *Phys. Rev.*, 184, 338, 1969.

38. Morgan, D. J. and Adams, M. J., Quantum noise in semiconductor lasers, *Phys. Status Solidi*, 11, 243, 1972.

39. Yamamoto, Y., AM and FM quantum noise in semiconductor lasers. I. Theoretical analysis, *IEEE J. Quantum Electron.*, 19, 34, 1983.

40. Yamamoto, Y., Saito, S., and Mukai, T., AM and FM quantum noise in semiconductor lasers. II. Comparison of theoretical and experimental results for AlGaAs lasers. *IEEE J. Quantum Electron.*, 19, 47, 1983.

41. Schimpe, R., Theory of intensity noise in semiconductor laser emision, *Z. Phys.* B-52, 289, 1983.

42. McCumber, D. E., Intensity fluctuations in the output of cw laser oscillators I, *Phys., Rev.*, 141, 306, 1966.

43. Ogawa, K. and Vodhanel, R. S., Measurements of mode partition noise of laser diodes, *IEEE J. Quantum Electron.*, 18, 1090, 1983.

44. Ogawa, K., Analysis of mode partition noise in laser transmission systems, *IEEE J. Quantum Electron.*, 18, 849, 1982.

45. Liu, P. and Ogawa, K., Statistical measurement as a way to study mode partition in injection lasers, *IEEE J. Lightwave Technol.*, 2, 44, 1984.

46. Ito, T., Machida, S., Nawata, K., and Ikegami, T., Intensity fluctuations in each longitudinal mode of a multi-mode AlGaAs-laser, *IEEE J. Quantum Electron.*, 13, 574, 1979.

47. Arnold, G. and Petermann, K., Intrinsic noise of semiconductor lasers in optical communication systems, *Optical Quantum Electron.*, 12, 207, 1980.

48. Koyama, F., Tabun-Ek, T., Arai, S., Wang, S., Suematsu, Y., and Furuya, K., Suppression of intensity fluctuations of a longitudinal mode in directly modulated GaInAsP/InP dynamic single mode lasers, *Electron. Lett.*, 19, 325, 1983.

49. Ogawa, K., Considerations for single-mode fiber systems, *Bell Syst. Technol. J.*, 61, 1919, 1982.

50. Grosskopf, G., Kueller, L., and Patzak, E., Laser mode partition noise in optical wideband transmission links, *Electron. Lett.*, 18, 493, 1981.

51. Goldberg, L., Taylor, H. F., Dandridge, A., Weller, J. F., and Miles, R. O., Spectral characteristics of semiconductor lasers with optical feedback, *IEEE J. Quantum Electron.*, 18, 555, 1982.

52. Kikuchi, K. and Okoshi, T., Simple formula giving spectrum-narrowing of semiconductor-laser output obtained by optical feedback, *Electron. Lett.*, 18, 10, 1982.

53. Wyatt, R. and Devlin, W. J., 10 kHz linewidth 1.5 μm InGaAsP external cavity laser with 55 nm tuning range, *Electron. Lett.*, 19, 110, 1983.

54. Olesen, H., Saito, S., Mukai, T., Saitoh, T., and Mikami, O., Solitary laser spectral linewidth and its reduction with external grating feedback for a 1.55 μm InGaAsP BH-laser, *Jpn. J. Appl. Phys.*, 22, L664, 1983.

55. Patzak, E., Sugimura, A., Saito, S., Mukai, T., Olesen, H., Semiconductor laser linewidth in optical feedback configurations, *Electron. Lett.*, 19, 1026, 1983.

56. Kuwahara, H., Imai, H., and Sasaki, M., Intensity noise of InGaAsP/InP lasers under the influence of reflection and modulation, *Opt. Commun.*, 49, 315, 1983.

57. Hirota, O. and Suematsu, Y., Noise properties of injection lasers due to reflected waves, *IEEE J. Quantum Electron.*, 15, 142, 1979.

58. Hirota, O., Suematsu, Y., and Kwok, K., Properties of intensity noises of laser diodes due to reflected waves from single-mode optical fibers and its reduction, *IEEE J. Quantum Electron.*, 17, 1014, 1981.

59. Ikushima, I. and Maeda, M., Self coupled phenoma of semiconductor lasers caused by an optical fiber, *IEEE J. Quantum Electron.*, 14, 331, 1978.

60. den Biesterbos, J. W. M., Boef, A. J., Linders, W., and Acket, G. A., Low frequency mode-hopping optical noise in AlGaAs channeled substrate laser induced by optical feedback, *IEEE J. Quantum Electron.*, 19, 986, 1983.

61. Ogasawara, N., Ito, R., Sasaki, T., Osada, T., and Suyama, M., Self-modulated light output from pulsed injection lasers with optical feedback, *Jpn. J. Appl. Phys.*, 21, L217, 1982.

62. Lang, R. and Kobayashi, K., External optical feedback effects on semiconductor injection laser properties, *IEEE J. Quantum Electron.*, 16, 347, 1983.

63. Osmundsen, J. H., Tromberg, B., and Olesen, H., Experimental investigation of stability properties for a semiconductor laser with optical feedback, *Electron. Lett.,* 19, 1068, 1983.

64. Ogasawara, N., Ito, R., Kato, M., and Takahashi, Y., Mode switching in injection lasers induced by temperature variation and optical feedback, *Jpn. J. Appl. Phys.,* 22, 1684, 1983.

65. Fujiwara, M., Kubota, K., and Lang, R., Low-frequency intensity fluctuations in laser diodes with external optical feedback, *Appl. Phys. Lett.,* 38, 217, 1981.

66. van der Ziel, J. P. and Mikulyak, R. M., Single mode operation of 1.3 μm InGaAsP/InP buried crescent lasers using short external optical cavity, *IEEE J. Quantum Electron.,* 20, 223, 1984.

67. Fleming, M. W. and Mooradian, A., Spectral characteristics of external cavity controlled semiconductor lasers, *IEEE J. Quantum Electron.,* 17, 44, 1981.

68. Hirose, Y. and Ogura, Y., Model for return beam induced noise generation in GaAlAs semiconductor lasers, *Electron. Lett.,* 16, 202, 1980.

69. Osmundsen, J. H. and Gade, N., Influence of optical feedback on laser frequency spectrum and threshold conditions, *IEEE J. Quantum Electron.,* 19, 465, 1983.

70. Saito, S., Nilsson, O., and Yamamoto, Y., Oscillation center frequency tuning, quantum FM noise and direct modulation characteristics in external grating loaded semiconductor lasers, *IEEE J. Quantum Electron.,* 18, 961, 1982.

71. Ichihashi, Y., Nagai, H., Miya, T., and Miyajima, Y., Transmission experiment over 134 km of single mode fiber at 445.8 Mb/s, IOOC 83, post deadline Technical Digest, 34, 1983.

72. Tsang, W. T., Olson, N. A., and Logan, R. A., High-speed direct single-frequency modulation with a large tuning rate and frequency excursion in cleaved-coupled-cavity semiconductor lasers, *Appl. Phys. Lett.,* 42, 650, 1983.

73. Toba, H., Kobayashi, Y., Yamagimoto, K., Nagai, H., and Nakahara, M., Injection-locking technique applied to a 170 km transmission experiment at 445.8 Mbit/s, *Electron. Lett.,* 20, 370, 1984.

74. Itaya, Y., Matsuoka, T., Kuroiwa, K., and Ikegami, T., Longitudinal mode behaviours of 1.5 μm range GaInAsP/InP distributed feedback lasers, *IEEE J. Quantum Electron.,* 20, 230, 1984.

75. Kobayashi, K., Utake, K., Abe, Y., and Suematsu, Y., Single-wavelength operation of 1.53 μm GaInAsP/InP buried-heterostructure integrated twin-guide laser with distributed Bragg reflector under direct modulation up to 1 GHz, *Electron. Lett.,* 17, 366, 1981.

76. Nakamura, M., Aiki, K., Umeda, J., and Yariv, A., Cw operation of distributed feedback GaAs-GaAlAs diode lasers at temperatures up to 300 K, *Appl. Phys. Lett.,* 27, 403, 1975.

77. Reinhart, F. K., Logan, R. A., and Shank, C. V., GaAs-Al$_x$Ga$_{1-x}$ As injection lasers with distributed reflectors, *Appl. Phys. Lett.,* 28, 596, 1976.

78. Utaka, K., Akiba, S., Sakai, K., and Matsushima, Y., Room temperature cw operation of distributed buried heterostructure InGaAsP/InP lasers emitting at 1.57 μm, *Electron. Lett.,* 17, 961, 1981.

79. Koyama, F., Suematsu, Y., Arai, S., and Tawee, T., 1.5—1.6 μm GaInAsP/InP dynamic-single-mode (DSM) lasers with distributed Bragg reflector, *IEEE J. Quantum Electron.,* 19, 1042, 1983.

80. Koyama, F., Arai, A., Suematsu, Y., and Kishino, K., Dynamic spectral width of rapidly modulated 1.58 μm GaInAsP/InP buried-heterostructure distributed-Bragg-reflector integrated-twin-guide lasers, *Electron Lett.,* 17, 938, 1981.

81. Coldren, L. A., Furuya, K., Miller, B. I., and Rentschler, J. A., Etched mirror and groove-coupled GaInAsP/InP laser devices for integrated optics, *IEEE J. Quantum Electron.,* 18, 1679, 1982.

82. Ito, M. and Kimura, T., Oscillation properties of AlGaAs DH lasers with external gratings, *IEEE J. Quantum Electron.,* 16, 69, 1980.

83. Chinone, N., Aiki, K., and Ito, R., Stabilization of semiconductor laser outputs by a mirror close to the facet, *Appl. Phys. Lett.,* 33, 990, 1978.

84. Preston, K. R., Woolard, K. C., and Cameron, K. H., External cavity controlled single longitudinal mode laser transmitter module, *Electron. Lett.,* 17, 931, 1981.

85. Elze, G., Grosskopf, G., Kueller, L., and Wenke, G., Experiments on modulation properties and optical feedback characteristics of laserdiodes stabilized by an external cavity or injection locking, *IEEE J. Lightwave Technol.,* LT-2, 1063, 1984.

86. Sato, H., Itoh, K., Fukai, M., and Suzuki, M., Single longitudinal mode control of semiconductor lasers by rectangular conical diffractor system for wavelength-division-multiplexing transmission, *IEEE J. Quantum Electron.,* 18, 328, 1982.

87. Sommers, H. S., Jr., Performance of injection lasers with external gratings, *RCA Rev.,* 38, 33, 1977.

88. Saito, S. and Yamamoto, Y., Direct observation of lorentzian lineshape of semiconductor laser and linewidth reduction with external grating feedback, *Electron Lett.,* 17, 325, 1981.

89. Lang, R., Injection locking properties of a semiconductor laser, *IEEE J. Quantum Electron.,* 18, 976, 1982.

90. Kobayashi, S. and Kimura, T., Injection locking in AlGaAs semiconductor laser, *IEEE J. Quantum Electron.,* 17, 681, 1981.

91. Kobayashi, S., Yamada, J., Machida, S., and Kimura, T., Single-mode operation of 500 Mbit/s modulated AlGaAs semiconductor laser by injection locking, *Electron. Lett.,* 16, 746, 1980.

92. Arnold, G., Petermann, K., Russer, P., and Berlec, F. J., Modulation behaviour of double heterostructure injection lasers with coherent light injection, *AEÜ*, 32, 129, 1978.
93. Lang, R. and Kobayashi, K., Suppression relaxation oscillation in the modulated output of semiconductor lasers, *IEEE J. Quantum Electron.*, 12, 520, 1977.

Chapter 5

# COUPLING EFFICIENCY AND OPTICAL FEEDBACK CHARACTERISTICS

Gerhard Elze and Gerhard Wenke

## TABLE OF CONTENTS

I.      Introduction...................................................................56
        A.      Coupling Efficiency in SM Fiber Systems....................56
        B.      Optical Feedback .................................................58

II.     Laser-Fiber Coupling .................................................59
        A.      Efficiency.............................................................59
        B.      Feedback..............................................................65

III.    Fiber Interconnection ...............................................66
        A.      Efficiency.............................................................66
        B.      Feedback..............................................................67

IV.     Fiber-Photodiode Coupling .......................................68
        A.      Efficiency.............................................................68
        B.      Feedback..............................................................68

V.      Optical Isolators .......................................................69

References ...............................................................................69

# I. INTRODUCTION

An essential factor in the implementation of an optical transmission system is the interconnection of the various components such as transmitters, fiber-optic transmission lines, repeaters, and receivers. From the point of view of mechanics, electrical connections are contact elements and optical connections are centering elements. In practice, the problem of coupling is not merely an optical problem, i.e., the problem of imaging and taking into account optical aberation and losses, it is just as much a problem of producing and adjusting sophisticated miniature components.

In the first part of this chapter, a description of optical coupling in single-mode (SM) fiber systems using Gaussian beam approximation will be given. In addition, a further important factor in optical wideband systems will be discussed, namely the influence of optical reflections at coupling points on the emission of the transmitter (see Chapter 4). This will be followed by an examination of laser-fiber coupling, fiber interconnection, and fiber-photodiode coupling, taking into account in each case the efficiency and the possible optical reflections (optical feedback). Treatment will also be given to techniques which are frequently employed to improve efficiency and reduce optical feedback.

## A. Coupling Efficiency in SM Fiber Systems

The coupling efficiency is treated in References 1 through 9. The efficiency for the coupling of light into a fiber is defined as follows:

$$\eta\ (\%) = \frac{P_{guided}}{P_{emitted}} \cdot 100 \tag{1}$$

where $P_{guided}$ = power of the light coupled into and guided along the fiber and $P_{emitted}$ = power of the light emitted from one side of the light source or from another fiber. The coupling loss is defined as:

$$\eta\ (dB) = -\ 10\ \log \frac{P_{guided}}{P_{emitted}} \tag{2}$$

The emitted power is the integral of the radiance over the emitting surface A and the solid angle $\Omega$.

The guided power is that part of the emitted light which is received and guided by the fiber depending on the numerical aperture of the fiber and the fiber core area.[2,7]

The coupling efficiency can be improved by increasing the numerical aperture and the core radius. However, enlargement of the aperture by raising the refractive index difference $\Delta n$ between the core and cladding (see Chapter 3) will bring about stronger doping and thus greater fiber losses due to scattering. Furthermore, in the case of SM fibers, the condition that the normalized fiber frequency V be <2.4 (see Chapter 3) must be fulfilled, thus restricting the dimensions of both the radius and aperture.

The optimum coupling efficiency can be obtained for identical field distribution of source and the fundamental mode of the fiber. The field distribution E(x) in SM fibers as well as in transversal SM semiconductor lasers can be described as having approximately a Gaussian profile. The error made is merely a few percent.[4,10,11]

$$E(x) = E(0)e^{-x^2/w_0^2} \tag{3}$$

where x = distance from beam axis, $w_0$ = radius of beam waist (spot size) for E(x) = E(0)/e at z = O. (The beam waist is the point where the field concentration of the Gaussian beam is highest.)

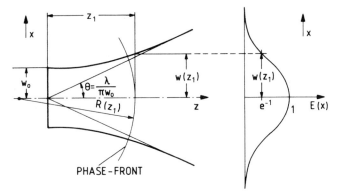

FIGURE 1. Gaussian beam with beam waist $w_0$ propagating in z-direction. Phase front $R(z_1)$ and variation of electrical field strength $E(x)$ at $z = z_1$ indicated.

The coupling of SM fibers both one to another and to lasers can therefore be regarded as constituting transformation and matching of Gaussian beams.

For large distances, the far-field angle $\theta_{1/e}$ of a Gaussian beam is derived from the spot size $w_0$ (see Figure 1):

$$\theta_{1/e} = \frac{\lambda}{\pi w_0} \tag{4}$$

Thus the spot size $w_0$ of a laser or a fiber can be determined by measuring the far-field. Usually the half-angle of intensity $\Theta_{1/2}$ of the laser is given. The relation to the far-field angle of a Gaussian beam is

$$\theta_{1/e} = \frac{\theta_{1/2}}{\sqrt{2 \ln 2}} \tag{5}$$

The spot size of a SM fiber can also be determined by fiber parameters:[10]

$$w_0 = a(0.65 + 1.619/V^{3/2} + 2.879/V^6) \tag{6}$$

where a = radius of fiber core and V = normalized frequency of fiber.

The relation of the radius of the beam waist $w_0$ at $z = 0$ to the radius of the beam $w(z)$ at location $(z)$ is as follows (see Figure 1):

$$w(z) = w_0\left(1 + \left[\frac{\lambda z}{\pi w_0^2}\right]^2\right)^{1/2} \tag{7}$$

The relation of $w_0$ to the radius of curvature $R(z)$ of the phase front at z is

$$R(z) = z\left(1 + \left[\frac{\pi w_0^2}{\lambda z}\right]^2\right) \tag{8}$$

Only for large distances to the beam waist (far-field) is the radius of curvature $R(z)$ identical to z.

Two Gaussian beams will exhibit optimum coupling when the spot sizes are matched. The easiest method is by imaging the beam waists into each other, i.e., transforming the spot size of the source by means of lenses to the location and to the dimensions of the spot size of the fiber at its front face.

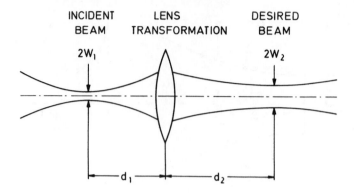

FIGURE 2.    Lens transformation of a Gaussian beam. The beam waist $w_1$ at $d_1$ is transformed into beam waist $w_2$ at $d_2$.

The transformation by means of an ideal lens having a focal length of f is defined as follows (see Figure 2):[4]

$$\frac{w_2}{w_1} = \frac{\sqrt{f^2 - f_0^2}}{d_1 - f} \qquad f \geq f_0 \tag{9}$$

where $f_0 = (\pi/\lambda) w_1 w_2$ and $d_1$ = distance of $w_1$ from the lens.

The power-coupling efficiency between Gaussian beams of spot size $w_1$ at location $d_1$ and spot size $w_2$ at location $d_2$ is as follows (see Figure 2):

$$\eta_{1/2} = \frac{4w_1^2 w_2^2}{(w_1^2 + w_2^2) + \lambda^2(d_1 + d_2)^2/\pi^2} \tag{10}$$

For the transformation of Gaussian beams through composed systems by means of beam matrices, and for the calculation of coupling losses with respect to misaligned and tilted beam waists, the reader is referred to special treatment of the subject.[8,10,12] The final calculation of coupling losses must also take into account Fresnel reflections (approximately 4% at the perpendicular quartz-to-air transition) and limitation of the Gaussian beam at boundaries.

## B. Optical Feedback

In operational optical lines, there are always some inhomogeneities at which light emitted by the laser is reflected or scattered, causing part of the light to return to the laser.

A number of typical reflection points are shown in Figure 3. These include the coupling arrangement and the front face of the fiber or coupling lenses. Fresnel reflections also occur in conventional optical connectors with an air gap. In unfavorable conditions even stronger reflections arise through interference. Light is backscattered from the fiber itself through inhomogeneities in the material. A considerable amount of light can also be reflected from a planar fiber end or from the photodiode. Generally speaking, one can expect greater optical feedback effects when SM fibers are used because of the strong coupling of the fiber to the laser for reflected light. The reflected light is not separated into various modes as is the case with multimode (MM) fibers. The effective intensity reflection factor $R_E$ of points of reflection in a SM fiber is expressed by:

$$R = \eta^2 \cdot e^{-2\alpha L} \cdot R_1 \cdot \frac{1}{2} \tag{11}$$

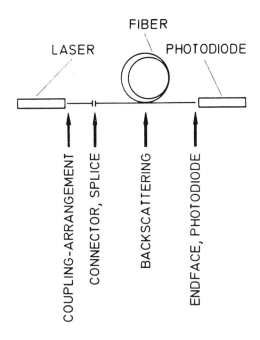

FIGURE 3.    Optical feedback in a fiber-optic transmission line. Arrows indicate typical causes such as Fresnel reflection and Rayleigh backscattering.

where $\eta$ = laser-fiber coupling efficiency, $\alpha$ = attenuation coefficient of the fiber, L = length of fiber up to the point of reflection, and $R_1$ = reflection factor at the point of reflection.

The factor 1/2 takes into account the existence of two polarization modes in the fiber. $R_E$ in Equation (11) varies depending on the distribution of the reflected light among the two modes.

In the case of index-guiding lasers (IGL), which are favored in wideband, long-distance systems, it has been estimated[13] that the reflection factor should not exceed $10^{-6}$ to $10^{-8}$ in order not to affect the laser performance. If the reflections are not sufficiently suppressed, the laser emission characteristics are changed.[32-41] The power output (laser threshold), the optical spectrum, the dynamic behavior and linearity undergo in some cases drastic changes. A distinction must be made between nearby reflections occurring within the coherence length of the laser and incoherent reflections from distances farther away (see Chapter 4 and References 32, 33, 35, and 37).

## II. LASER-FIBER COUPLING

### A. Efficiency

As has already been mentioned, the relations for matching Gaussian beams are also valid for coupling lasers to SM fibers. For most available lasers, the light-emitting spot is not circular but elliptical, i.e., in terms of Gaussian beams, different beam waists are to be considered perpendicular (x) and parallel (y) to the junction plane.

$$w_{0y} > w_{0x} \qquad \theta_y < \theta_x \qquad \text{from (4)} \qquad (12)$$

An approximate assessment of losses can be arrived at using the geometrical mean

$$w_0 = \sqrt{w_{0y} \cdot w_{0x}} \qquad (13)$$

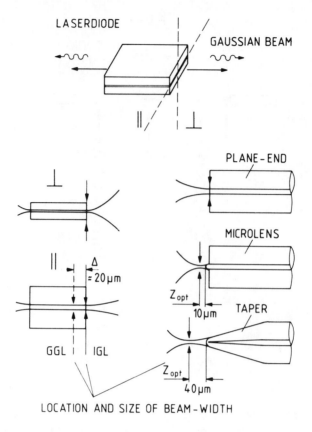

FIGURE 4.     Location and size of the Gaussian beam waist for an IGL and a GGL. The beam waist for the parallel direction of GGL is approximately 20 $\mu$m inside the active zone. The beam waist for three differently prepared fiber coupling faces is depicted on the right.

With regard to gain-guided lasers (GGL), it must also be noted that the beam waist parallel (y) to the junction plane is situated within the laser and not at the exit mirror (see Figure 1 and 4).

$$z_{0x} \neq z_{0y} \tag{14}$$

This astigmatic behavior of GGL is due to the absence of index-guiding (see Chapter 4) in the junction plane (y), resulting in a bending of the phase front.[9]

The shift as expressed by $z_{0x} - z_{0y} = \Delta$ is given as amounting to 20 $\mu$m.[14] In the same reference, a description is given of how even in the event of astigmatism, high coupling efficiency can be achieved by using three lenses, including the virtual imaging of a cylindrical lens.[57,58] Because of their more favorable spectral behavior, IGL are given preference in high bit rate and long-distance systems. Astigmatism in this case is either completely absent or negligibly small. In some types, there is also hardly any ellipticity, e.g., in BH (buried heterostructure) lasers. The coupling efficiency into fibers, especially for these types of lasers, is greatly improved by shaping the fiber front face into a microlens[12,30,31] which acts as a phase object. Such microlenses enable the Gaussian beam of the laser to be approximately matched to the parallel wavefronts in SM fibers. Some examples[29] should be mentioned at this point. It is reported that after heating and applying pressure (between two plates) to a special fiber with low core viscosity, some of the core material is squeezed out, thus forming the microlens.[15] Another

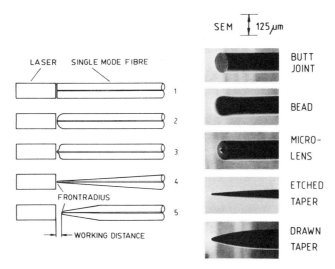

FIGURE 5. Widely used fiber coupling faces investigated in Reference 20. Left: five different principles for direct coupling; right: scanning electron micrographs (SEM) of implemented fiber faces. (1) Cleaved butt joint; (2) flame-polished hemisphere; (3) microlens (small amount of deposited silica); (4) taper (cladding etched off); (5) taper (drawn in an electric arc). The drawn taper has a large working distance of 30 to 50 $\mu$m for high coupling efficiency.

method involves the application of photoresist to the fiber front face which is exposed by light coupled into the opposite end of the fiber.[11] After developing, a lens-shaped profile is left which centers automatically on the core. Selective etching of the core and cladding constitutes a further method of self-centering.[16,17,18] This procedure makes use of the different etching speeds of the core and cladding materials which are doped differently. The result is either a peak of the core material which has the effect of a lens, or a dip at the core region in which a microlens can be positioned. It is also possible to polish the fiber end into a quadrangular pyramid and afterwards round off the tip in the form of a lens in an electric arc.[19] This method permits various lens radii to be used both perpendicular and parallel to the junction plane of the laser. A further technique, besides selective etching, is that of etching off the cladding material and melting the remaining tip in an electric arc to a small sphere which is centered to the cladding but not necessarily on the core.[20] By heating and drawing the fiber in an electric arc, both the core and cladding are tapered. Subsequently, the tip is melted into a microlens.[20-23] Such a taper is easy to make and is easily reproduced. There also exist modified versions with a high-index microlens on top of the tapered region.[24]

In most cases the working distance for optimum coupling of a microlensed fiber face can be estimated by a simple relation for the focal length:

$$f = r_F/(n_F - 1) \tag{15}$$

where $r_F$ = front radius of the lens and $n_F$ = fiber (lens) refractive index.

In Reference 20 there is a comparison of several methods of coupling, all of which are easily produced (Figure 5).

Coupling efficiencies amounting to approximately 2 dB can be obtained employing channeled substrate planar (CSP) lasers. Coupling efficiencies and displacement tolerances for the respective methods are depicted in Figure 6 and summarized in Table 1.

The optimization of one of the structures, a drawn taper produced by successively enlarging the front radius, is illustrated in Figure 7.

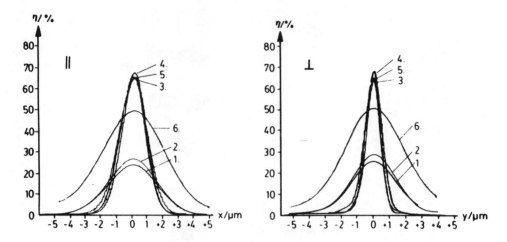

FIGURE 6.    Left: coupling efficiency and displacement tolerance x parallel to the junction plane; right: same for y perpendicular to the junction plane. The laser used is a Hitachi® HLP 1400 CSP laser. 1 through 5 are described in Figure 5. 6 corresponds to 2 Selfoc® lenses and a plane fiber end. The coupling efficiencies are not corrected to the laser power change under the respective coupling conditions.

## Table 1
### RESULTS OF COUPLING EFFICIENCY AND DISPLACEMENT TOLERANCE FOR COUPLING STRUCTURES DESCRIBED IN FIGURES 5 AND 6

| Coupling arrangement (fiber-endface) | Optimum distance z or safety distance (μm) | Coupling efficiency (%) | Coupling loss (dB) | Tolerance (−1 dB) (μm) | | |
|---|---|---|---|---|---|---|
| | | | | x | y | z |
| Plane-end | 10 | 21 | 6.9 | ±1.1 | ±0.9 | +8.7 |
| Bead | 5—10 | 22.5 | 6.5 | ±1.0 | ±0.9 | +7.9 |
| Microlens | 10 | 59 | 2.3 | ±0.6 | ±0.35 | ±4.2 |
| Etched taper | 5—10 | 63 | 2.2 | ±0.6 | ±0.4 | ±4 |
| Drawn taper | 30—50 | 66 | 1.8 | ±0.7 | ±0.4 | ±6.6 |
| 2 Selfoc® lenses + plane-end | ≅300 | 46 | 3.4 | ±1.4 | ±1.3 | ⩾ 20 |

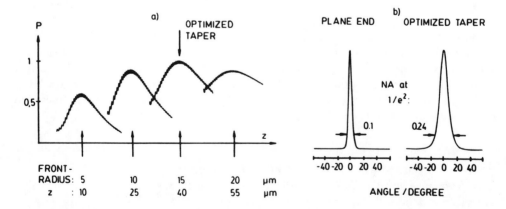

FIGURE 7.    (a) Optimization of a drawn taper by successive enlargement of the front end radius; (b) far field emitted from a plane-end (butt joint) and a drawn taper. The light is launched into the opposite side of a 3.5-km long fiber.

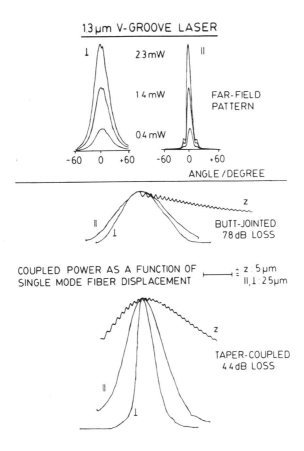

FIGURE 8.    Far-field pattern at various power emissions for
a 1.3-μm V-groove laser. Coupled power for a butt joint and a
drawn taper at about 1-mW laser power.

A taper of length approximately 300 μm and front radius 15 to 20 μm has a considerably enlarged numerical aperture (NA) compared to that of a plane end (Figure 7b).
The NA is measured by investigating the far-field emission of the variously prepared
fiber ends.

Calculation of the coupling efficiency with the relations for Gaussian beams (4), (5),
and (10) from the measured far-fields of laser and fiber (plane ends and taper) gives
the following results:

- $\eta$    plane    = 0.4      (calculated)
- $\eta$    plane    = 0.21    (measured)
- $\eta$    taper    = 0.8      (calculated)
- $\eta$    taper    = 0.66    (measured)

The differences between the values measured and those calculated are mainly due to
the approximation of the elliptical spot of the CSP laser ($w_{0x} = 0.66$ μm, $w_{0y} = 1.62$
μm) using the geometric mean.

Figures 8 and 9 provide examples of the coupling conditions for long wavelength
lasers (1.3 μm BH laser and 1.28 μm V-groove laser), i.e., the far-field patterns and
displacement tolerances for plane fiber ends and tapers in a coupled state.

It is also possible to obtain good coupling results with Selfoc® lenses.[10,20,25] Such
graded-index rod lenses can be favorably used when employing optical isolators or
wavelength division multiplexers. The coupling arrangements with Selfoc® lenses for
the laboratory system described in Chapter 10 (system 2) are depicted in Figure 10.

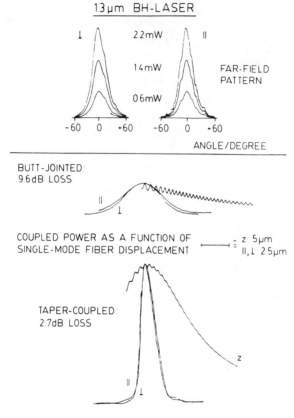

FIGURE 9.     Far-field pattern at various power emissions for
a 1.3-μm BH laser. Coupled power for a butt joint and a drawn
taper at about 1-mW laser power.

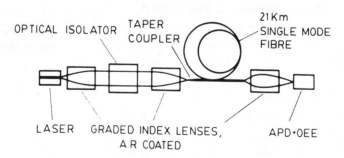

FIGURE 10.     Transmission line with optical isolation (system 2; see
Chapter 10).

From Figure 6 it can be seen that in the case of laser-fiber coupling with two Selfoc® lenses, the coupling efficiency is slightly reduced compared to direct coupling with microlensed fiber faces, but the lateral displacement tolerances are much larger.

A report on investigations into the displacement tolerances of coupling methods using a combination of two lenses in confocal conditions is provided in Reference 26; coupling into lithium niobate waveguides is treated extensively in References 62 and 63.

As a general rule, the laser-fiber coupling should be hermetically sealed for nonlaboratory transmission systems so that it is protected against moisture and contamination. References 27 and 28 contain examples of protective encapsulation.

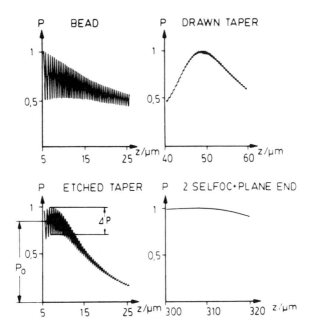

FIGURE 11. Power changes on coupling into three different fiber faces (Figure 5) and on using Selfoc® lenses (Figure 10). The power is measured behind a 3.5-km fiber as a function of the laser-fiber distance z.

## B. Feedback

The effect of "near" reflections from the fiber face and from coupling lenses on the laser emission characteristics is described with the Compound Cavity Model[32,33,35,40] (see Chapter 4). Usually the direction of polarization is not altered by reflections of this kind. The reflections occur within the coherence length of the laser. Because of this, a strong influence on the laser performance can be observed. It depends on the phase of the reflected light. Typical periodic variations in the power-output $\Delta P$ (laser threshold) of a CSP laser when the distance between the laser and a coupling fiber is changed, are depicted in Figure 11.[20]

The measurements are taken for various coupling methods. A taper, drawn in an electric arc and provided with a small hemispherical end face (Figures 4 and 5), causes less power variation near the optimum working distance of 30 to 50 $\mu$m compared with, e.g., a hemispherical fiber end or an etched taper with a microlens having working distances of <10 $\mu$m for high coupling efficiency. In Reference 23, reflections of −50 to −53 dB at optimum distance are reported for taper structures compared with −25 to −27 dB for plane fiber ends at a distance of approximately 5 $\mu$m to the laser facet. With Selfoc® lenses where working distances of approximately 300 $\mu$m as well as antireflection coating are normal, the power variation is even less (Figure 11).

In Chapter 4 a description is provided of the effect of a nearby reflector, e.g., the fiber front face, on the optical spectrum. Changes in laser-fiber coupling distances are accompanied by periodic wavelength variations $\Delta\lambda$. The results of measurements of the influence of various coupling techniques on the laser spectrum are reported.[20] In Figure 12 the wavelength variations are registered near the respective optimum working distances.

The influence on the laser spectrum is described by the feedback parameter X (Chapter 4) containing the reflectivity *and* the distance to the reflection point. The influence on the laser threshold (power output) is a function of the reflectivity alone, i.e., the amount of light reflected back into the laser.[42]

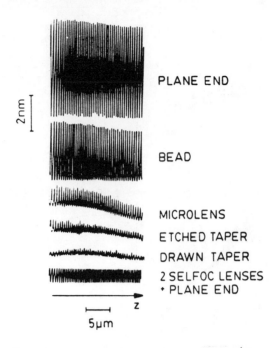

FIGURE 12.    Laser wavelength variation as a function
of the distance between the CSP laser facet and the particular coupling methods; e.g., the large wavelength
variation for the plane-end corresponds to 16 mode
spacings. For details, refer to Reference 20.

Since the stability of a SM laser spectrum is an essential requirement for wideband
optical transmission with large repeater distances (Chapter 10), methods for laser-to-
fiber couplings should be compared not only by their coupling efficiency $\eta$ or their
influence on the laser threshold but also by their influence on the laser spectrum. In
Chapter 4 methods are discussed showing how reflectors can be used to select laser
modes, permitting SM operation of a laser which without selection would operate in
several modes.

Figure 13 shows a transmission unit containing a long wavelength BH laser, a stable
external reflector at the rear of the laser, and taper coupling at the laser front facet.

## III. FIBER INTERCONNECTION

### A. Efficiency

The coupling of SM fibers by splicing or by means of optical connectors has been
given extensive treatment.[1,2,3,43,44,49] For nondetachable connections, techniques such
as melting or gluing the fiber ends together are applied. The fiber front ends are carefully cleaved and placed in position, usually in the case of SM fibers, by observing and
optimizing the amount of coupled light. This demands that the light be measured at
the far end of the fiber or that the light coupled into the cladding be observed and
minimized.[29] Splicing can also be optimized by use of an interference contrast micro-
scope.[46] The fiber ends are then either welded in an electric arc or glued with UV-
curable cement.[45] For shifted zero dispersion fibers, gluing is to be preferred since it
prevents additional losses due to core deformation on heating of the fiber.[47] Equipment
is also available for semiautomatic SM fiber splicing. As a final step the splice is provided with a sleeve for protection.

FIGURE 13.    Laser chip with an external reflector at a distance of 500 μm from the laser back facet. (The optimum distance from the best sidemode suppression is about 200 μm.) The light is launched into a tapered fiber end at the laser front facet.

If detachable fiber connections are required, optical connectors are employed. Various approaches exist for designing such connectors. The end face coupling method requires high-accuracy core alignment, a typical value being < 1 μm. Methods of alignment and production of low-loss, SM connectors are described.[48,61]

Another method is the lensed connector (expanded beam type), exhibiting higher alignment tolerances. This has good reproducibility but compared to the endface connector it generally also has higher coupling losses. The losses are due to Fresnel reflections at several glass-to-air transitions and in some cases poor lens imaging quality. Index matching fluid to reduce Fresnel losses can be used in the endface connector but not in the expanded beam lensed connector. The problem of aging and contamination of index matching fluids has not yet been solved. As regards optical losses for long-distance transmission, high-accuracy endface type connectors would appear to be most suitable. However, for applications where loss is of less concern but high reproducibility is necessary, the lensed connector is more suitable. High quality connectors of both types are commercially available.

## B. Feedback

As a result of reflection of optical connectors with an air gap (typical value, 2 μm) at a distance 1 of about 0.5 to 2 m from the laser, i.e., within the coherence length of the light, the laser mode divides into several submodes spaced by $\Delta v$:

$$\Delta v = \frac{1}{\tau} \text{ equivalent to } \Delta \lambda = \frac{\lambda^2}{c \cdot \tau} \tag{16}$$

where $\tau$ = round trip time laser-to-connector.

$$(1 \text{ m connector: } \Delta v = \frac{1}{\tau} = \frac{c}{2 \text{ nL}} \cong 100 \text{ MHz})$$

These submodes are often unstable, owing to unstable reflection conditions arising from mechanical vibrations and microphony. They are, in part, mode-locked[50] and give rise to pronounced noise.

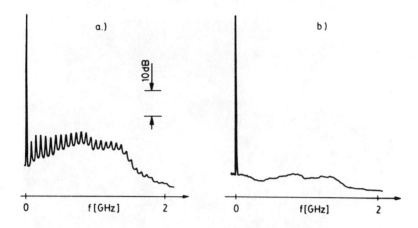

FIGURE 14.    Power density noise spectra in the baseband measured after 21-km fiber with an optical connection at a distance of 1.4 m from the laser. (a) Conventional connector with air gap; (b) low reflection connector; endfaces have physical contact.

The effects of connector reflection in the baseband are shown in Figure 14 (see also Chapter 10). Periodic, extremely pronounced noise enhancement occurs spaced by $\Delta v$.

Low frequency instabilities of laser diodes with optical feedback are described.[51] These fluctuations with period of approximately 100 nsec can be explained by shifts in the laser refractive index in the case of optical reflections.

By using index-matching fluid, the Fresnel reflections and interference effects can be suppressed. A better solution involves the use of optical connectors whose fiber ends establish physical contact.[44] The feedback effects due to Rayleigh backscattering from an optical fiber are described.[64]

## IV. FIBER-PHOTODIODE COUPLING

### A. Efficiency

Due to the large diameter of the photosensitive area of the photodiode — usually about 15 to 100 $\mu$m for high-speed detectors — the coupling of the fiber and photodiode is not critical. Using microlenses and applying AR-coating, it is possible to achieve coupling efficiencies of almost 100%. A common method is that of gluing the fiber directly on to the protective window of the photodiode. If an optical connector is used at the receiver end, it can be positioned immediately in front of the diode.

### B. Feedback

Effects on the laser emission characteristics of reflections from distances beyond the coherence length of the laser are described with the aid of the Injection State Model[32,33,35] (see Chapter 4). It should be borne in mind however, that if submodes arise (see above), they will have a very large coherence length due to their narrow spectra and will therefore reflect coherently. The light, which is usually subject to substantial fluctuations in both amplitude and phase, returns to the laser, is amplified and reemitted. This leads to considerable noise enhancement as well as to changes in the linearity of the laser.[37,41]

As regards optical transmission systems, the basic difference between near and far reflections lies in the fact that far reflections, due to their long round trip time, cause noise enhancement in the laser itself even without the presence of a dispersive optical fiber. Near reflections on the other hand, alter the laser working point as well as, on some occasions, the modulation performance, and can change SM operation into un-

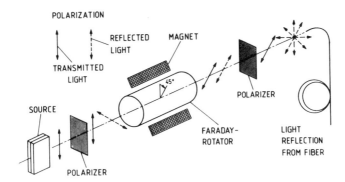

FIGURE 15.    Principle of an optical isolator, magneto-optic type.
Reflected light of any polarization is suppressed by typical factor of
20 to 25 dB. Insertion losses are 1 to 1.5 dB.

stable MM behavior. The main distortions occur in this case as partition noise (see Chapters 4 and 10) when dispersive optical fiber is used for transmission.

## V. OPTICAL ISOLATORS

Reflections from coupling arrangements in the immediate vicinity of the laser mirror can be reduced by choosing a greater working distance, divergent reflection, and if possible, by application of AR-coating. The remaining reflections can be suppressed by using optical isolators. A polarizer and a λ/4 plate is suitable for isolation of a specific wavelength. A polarizer and a Fresnel rhomb retarder is usable for a variety of wavelengths by altering the angle of incidence into the Fresnel rhomb. However, this type of isolator is not capable of suppressing reflected light completely if changes in the state of polarization occur due to optical fiber influences.

A technique which can be applied to suppress light of all states of polarization involves the use of a Faraday rotator in connection with two polarizers (Figure 15).

The Faraday material is usually a rare earth for the optical short wavelength range or YIG[56] for the optical long wavelength range. Such components are commercially available[52] and exhibit a typical insertion loss of 1 to 1.5 dB and an isolation of 20 to 25 dB. In Reference 53 there is mention of a relatively small sized isolator which can be integrated into an optical cable or connector. A toroidal configuration using the fiber itself as a Faraday rotator has also been implemented in practice.[54,59,60] In addition, it is possible to use Faraday rotating material taking the form of a small sphere in conjunction with a magnet for laser-to-fiber coupling. This not only has the effect of improving the coupling efficiency but also serves as isolation.[55] It would appear that the problem of monolithic integration of a transmitting laser and an effective isolator has not yet been completely solved.

## REFERENCES

1. Unger, H. G., *Planar Optical Waveguides and Fibres,* Clarendon Press, Oxford, 1977.
2. Wolf, H. F., *Handbook of Fiber Optics,* Garland STPM Press, New York, 1979.
3. Midwinter, J. E., *Optical Fibers for Transmission,* John Wiley & Sons, New York, 1979.
4. Marcuse, D., *Light Transmission Optics,* (Bell Laboratories Series), Van Nostrand Reinhold, New York, 1972.
5. Grau, G., *Optische Nachrichtentechnik,* Springer-Verlag, Berlin, 1981.

6. Unger, H. G., *Optische Nachrichtentechnik,* Elitera-Verlag, Berlin, 1976.
7. Wagner, R. E. and Tomlinson, W. J., Coupling efficiency of optics in single-mode fiber components, *Appl. Opt.,* 21, 2671, 1982.
8. Lecoy, P. and Richter, H., Berechnung der Transmission von elliptischen Mikrolinsen zur Optimierung der Kopplung zwischen Halbleiterlaser und Monomodefaser, Technischer Bericht 455 TBR 72, Deutsche Bundespost, 1981.
9. Nicia, A., Loss analysis of laser-fiber coupling and fiber combiner and its application to wavelength division multiplexing, *Appl. Opt.,* 21, 4280, 1982.
10. Saruwatari, M. and Nawata, K., Semiconductor laser to single-mode fiber coupler, *Appl. Opt.,* 18, 1847, 1979.
11. Bear, P. D., Microlenses for coupling single-mode fibers to single-mode thin-film waveguides, *Appl. Opt.,* 19, 2906, 1980.
12. Sakai, J. I. and Kimura, T., Design of a miniature lens for semiconductor laser to single-mode fiber coupling, *IEEE J. Quantum Electron.,* 16, 1059, 1980.
13. Petermann, K. and Arnold, G., Noise and distortion characteristics of semiconductor lasers in optical fiber communication systems, *IEEE J. Quantum Electron.,* 18, 543, 1982.
14. Krumpholz, O. and Westermann, F., Power coupling between Monomode Fibres and Semiconductor Lasers with Strong Astigmatism, Paper 7,7-1, in Proc. 7th ECOC, Copenhagen, 1981.
15. Khoe, G. D., New coupling techniques for single-mode optical fibre transmission systems, Paper 6.1-1, in Proc. 5th ECOC, Amsterdam, 1979.
16. Eisenstein, G. and Vitello, D., Chemically etched conical microlenses for coupling single-mode lasers into single-mode fibers, *Appl. Opt.,* 21, 3470, 1982.
17. Hopland, S. and Berg, A., Fabrication of coupling fibres with spherical end faces by a selective etching/melting technique, Paper 9.3-1, in Proc. 5th ECOC, Amsterdam, 1979.
18. Kayoun, P., Puech, C., Papuchon, M., and Arditty, H. J., Improved coupling between laser diode and single-mode fibre tipped with a chemically-etched self-centred diffracting element, *Electron. Lett.,* 17, 401, 1981.
19. Sakaguchi, H., Seki, N., and Yamamoto, S., Power coupling from laser diodes into single-mode fibres with quandrangular pyramid-shaped hemiellipsoidal ends, *Electron. Lett.,* 17, 426, 1981.
20. Wenke, G. and Zhu, Y., Comparison of efficiency and feedback characteristics of techniques for coupling semiconductor lasers to single-mode fiber, *Appl. Opt.,* 22, 3837, 1983.
21. Kuwahara, H., Sasaki, N., Tokoyo, N., Saruwatari, M., and Nakagawa, K., Efficient and reflection-insensitive coupling from semiconductor lasers into tapered hemispherical-end single-mode fibres, Proc. 6th ECOC, Session 7, York, 1980, 191.
22. Kuwahara, H., Sasaki, M., and Tokoyo, N., Efficient coupling from semiconductor lasers into single-mode fibres with tapered hemispherical ends, *Appl. Opt.,* 19, 2578, 1980.
23. Kuwahara, H., Onoda, Y., Goto, M., and Nakagami, T., Reflected light in the coupling of semiconductor lasers with tapered hemispherical end fibers, *Appl. Opt.,* 22, 2732, 1983.
24. Khoe, G. D., Poulissen, J., and de Vrieze, H. M., Efficient coupling of laser diodes to tapered monomode fibres with high-index end, *Electron. Lett.,* 19, 206, 1983.
25. Nippon Sheet Glass, *Selfoc Handbook,* NSG America, Clark, NJ.
26. Saruwatari, M. and Sugie, T., Efficient laser diode to single-mode fiber coupling using a combination of two lenses in confocal condition, *IEEE J. Quantum Electron.,* 17, 1021, 1981.
27. Tachikawa, Y. and Saruwatari, M., Design and performance of metal-sealed laser-diode coupler for optical subscriber transmission, Paper 30C2-1, in Proc. IOOC, Tokyo, 1983.
28. Saruwatari, M. and Sugie, T., Highly efficient and stable laser-diode coupler for single-mode fiber transmission, Paper 27C2-6, in Proc. IOOC, Tokyo, 1983.
29. Khoe, G. D., Advanced passive componentry for optical fiber communication systems, Paper 29C3-3, in Proc. IOOC, Tokyo, 1983.
30. Murakami, Y., Yamada, J. I., Sakai, J. I., and Kimura, T., Microlens tipped on a single-mode fibre end for InGaAsP laser coupling improvement, *Electron. Lett.,* 16, 321, 1980.
31. Yamada, J. I., Murakami, Y., Sakai, J. I., and Kimura, T., Characteristics of a hemispherical microlens for coupling between a semiconductor laser and single-mode fiber, *IEEE J. Quantum Electron.,* 16, 1067, 1980.
32. Lang, R. and Kobayashi, K., External optical feedback effects on semiconductor injection laser properties, *IEEE J. Quantum Electron.,* 16, 347, 1980.
33. Hirota, O. and Suematsu, Y., Noise properties of injection lasers due to reflected waves, *IEEE J. Quantum Electron.,* 15, 142, 1979.
34. Hirota, O., Suematsu, Y., and Kwok, K. S., Properties of intensity noises of laser diodes due to reflected waves from single-mode optical fibers and its reduction, *IEEE J. Quantum Electron.,* 17, 1014, 1981.
35. Kanada, T. and Nawata, K., Injection laser characteristics due to reflected optical power, *IEEE J. Quantum Electron.,* 15, 559, 1979.

36. Fujiwara, M., Kubota, K., and Lang, R., Low-frequency intensity fluctuation in laser diodes with external optical feedback, *Appl. Phys. Lett.*, 38, 217, 1981.

37. Wenke, G. and Elze, G., Investigation of optical feedback effects on laser-diodes in broadband optical transmission systems, *J. Opt. Commun.*, 2, 128, 1981.

38. Goldberg, L., Taylor, H. F., Dandridge, A., Weller, J. F., and Miles, R. O., Spectral characteristics of semiconductor lasers with optical feedback, *IEEE J. Quantum Electron.*, 18, 555, 1982.

39. Osmundsen, J. H. and Gade, N., Influence of optical feedback on laser frequency spectrum and threshold conditions, *IEEE J. Quantum Electron.*, 19, 465, 1983.

40. Salathé, R. P., Diode lasers coupled to external resonators, *Appl. Phys.*, 20, 1, 1979.

41. Kikushima, K., Hirota, O., Shindo, M., Stoykov, V., and Suematsu, Y., Properties of harmonic distortion of laser diodes with reflected waves, *J. Opt. Commun.*, 3, 129, 1982.

42. Bludau, W. and Rossberg, R., Characterization of laser-to-fibre coupling techniques by their optical feedback, *Appl. Opt.*, 21, 1933, 1982.

43. Siemens Telcom Report 6, Special issue: "Nachrichtenübertragung mit Licht", 1983.

44. Suzuki, N. and Nagano, O., Low insertion and high return loss optical connectors for use in analog video transmission, Paper 30A3-5, in Proc. IOOC, Tokyo, 1983.

45. Miller, C. M., Deveau, G. F., and Smith, M. Y., Low loss single mode fiber splices using ultraviolet-curable cement, Paper 30A3-6, in Proc. IOOC, Tokyo, 1983.

46. Matsumoto, M., Haibara, T., Tanifuji, T., and Tokuda, M., A new monitoring method for axis alignment of single-mode optical fibre and splice loss estimation, Paper 30A3-7, in Proc. IOOC, Tokyo, 1983.

47. McCartney, D. J., Payne, D. B., and Wright, J. V., Analysis of splices in shifted zero dispersion monomode fiber, *Electron. Lett.*, 20, 78, 1984.

48. Khoe, G. D., van Leest, J. H. F. M., and Luijendijk, J. A., Single-mode fiber connector using core-centered ferrules, *IEEE J. Quantum Electron.*, 18, 1573, 1982.

49. Marcuse, D., Loss analysis of single-mode fiber splices, *Bell Systems Tech. J.*, 56, 703, 1977.

50. Ikushima, I. and Maeda, M., Lasing spectra of semiconductor lasers coupled to an optical fiber, *IEEE J. Quantum Electron.*, 15, 844, 1979.

51. Ries, R. and Sporleder, F., Low frequency instabilities of laser diodes with optical feedback, Paper BIV-3, in Proc. 8th ECOC, Cannes, 1982.

52. Nippon Electric Company, Data Sheet, Optical Isolators, 1983.

53. Shirasaki, M. and Asama, K., Compact optical isolator for fibers using birefringent wedges, *Appl. Opt.*, 21, 4296, 1982.

54. Findakly, T., Single-mode fiber isolator in toroidal configuration, *Appl. Opt.*, 20, 3989, 1981.

55. Sugie, T. and Saruwatari, M., Non-reciprocal circuit for laser-diode to single-mode fibre coupling employing a yig sphere, *Electron. Lett.*, 18, 1028, 1982.

56. Kobayashi, K., Matsushita, S., Seki, M., Odagiri, Y., Mito, I., and Sugimoto, M., Stabilized 1.3 μm laser diode-isolator module for a hybrid optical integrated circuit, Proc. Top. Meet. Integrated Guided Wave Opt., January 28 to 30, Asilomar, 1980.

57. Weidel, E., New coupling method for GaAs-laser-fibre coupling, *Electron. Lett.*, 11, 437, 1975.

58. Weidel, E., Light coupling from a junction laser into a monomode fibre with a glass cylindrical lens on the fibre end, *Opt. Commun.*, 12, 93, 1974.

59. Kawakami, S. and Shiraishi, K., Components for fiber-type isolator, Paper 29C3-1, in Proc. IOOC, Tokyo, 1983.

60. Miyazaki, Y. and Kagami, M., Waveguide type magneto-optic devices using bi-garnet thin films for modulation and isolation, Paper 29C3-2, in Proc. IOOC, Tokyo, 1983.

61. Khoe, G. D., van Leest, J., and Luijendijk, J. A., Single-mode fiber connector using core-centered ferrules, *IEEE J. Quantum Electron*, 18, 1573, 1982.

62. Alferness, R. C., Ramaswamy, V. R., Korotky, S. K., Divino, M. D., and Buhl, L. L., Efficient single-mode fibre to titanium diffused lithium niobate waveguide coupling for $\lambda = 1.32$ μm, *IEEE J. Quantum Electron.*, 18, 1807, 1982.

63. Hammer, J. M. and Neil, C. C., Observations and theory of high-power butt coupling to LiNbO₃-type waveguides, *IEEE J. Quantum Electron.*, 18, 1751, 1982.

64. Goldberg, L. Taylor, H. F., and Weller, J. F., Feedback effects in a laser diode due to rayleigh backscattering from an optical fibre, *Electron. Lett.*, 18, 354, 1982.

Chapter 6

# OPTICAL DETECTORS

Wolfgang Albrecht, Gerhard Elze, and Gerd H. Grosskopf

## TABLE OF CONTENTS

I.    Introduction.................................................................................74

II.   Performance Considerations................................................................74
      A.    PIN Photodiodes ...................................................................74
      B.    Avalanche Photodiodes (APD)......................................................76
      C.    Equivalent Circuit Diagram .......................................................77
      D.    Photodiode Materials for the Optical Long Wavelength Range .........78

III.  Properties of Photodiodes in BB Transmission Links ...........................79
      A.    Bandwidths of Photodiodes.......................................................79
      B.    Noise Considerations in APDs ...................................................80
            1.    Noise Sources.............................................................80
                  a.    Photocurrent Noise.............................................81
                  b.    Dark Current Noise .............................................83
            2.    Noise Properties of Ge-APDs.........................................84

References ...................................................................................84

## I. INTRODUCTION

This chapter presents a number of properties of photodiodes which are of particular importance for use in optical broadband (BB) transmission systems. It is not our intention to cover the principles of photodetection in detail in this book. However, these aspects can be found in other publications.[1-3]

Photodiodes convert the light signals received from the fibers into electrical signals (Chapter 2, Figure 1). In optical BB transmission links the following properties of the photodetectors are important:

- Long lifetime
- Large bandwidth
- Low noise
- High quantum efficiency
- Low dark current

Optical transmission technology only uses photodetectors constructed on a semiconductor basis. We distinguish between two types of detectors:

- PIN photodiodes
- Avalanche photodiodes (APD)

The main feature of the latter is that the avalanche effect produces an internal gain of the photocurrent.

## II. PERFORMANCE CONSIDERATIONS

### A. PIN Photodiodes

Figure 1a shows the cross-sectional view. A high-resistance intrinsic layer exists between the p- and n-layers. Photons are radiated to this layer through the p-layer. If the energy of the photons ($E_P = h \cdot v$) is greater than the band gap of the semiconductor material, this produces pairs of charge carriers. In normal operation the PIN photodiode is reverse biased. Therefore the carriers travel to the electrodes due to the electric field in the depletion region. The generation of the photocurrent is demonstrated in Figures 1b and 1c using a simple energy band model. Due to the reflection factor r on the surface of the diode, only the component $P = Po (1 - r)$ penetrates the diode. In the absorbing region light power declines in accordance with exp ($-\alpha_D \cdot x$). The absorption coefficient $\alpha_D$ or the depth of penetration $x_0 = 1/\alpha_D$ are characteristics of the semiconductor material and are dependent on the energy of the photons, i.e., on the light wavelength. Figure 2 shows the absorption coefficients $\alpha_D$ or depth of penetration $x_0$, respectively, for a number of semiconductor materials which are used for photodetectors.[4] The quantum efficiency $\eta$ of a photodiode is the ratio of the number $n_L$ of charge carriers produced, compared with the number of irradiated photons $n_{ph}$:

$$\eta = \frac{n_L}{n_{ph}} \tag{1}$$

The performance of a photodiode is often characterized by the responsivity R, which is related to the quantum efficiency by

$$R = \eta \frac{e}{h \cdot v}$$

e = Elementary charge $\tag{2}$

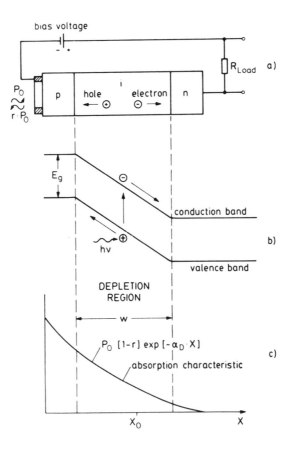

FIGURE 1.   Cross-sectional view of a PIN photodiode and corresponding simple energy band diagram.

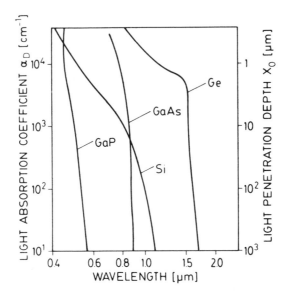

FIGURE 2.   Optical absorption coefficient for several semiconductor materials.

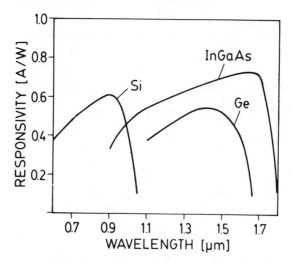

FIGURE 3.    Spectral responsivity of a PIN photodiode for different semiconductor materials.

In Figure 3 the responsivity R of different diode materials is shown as a function of the light wavelength.[5] At the levels usually used in optical transmission links there is a linear relationship between the received light power and the photocurrent:

$$i_{Ph} = \eta \frac{e}{h \cdot \nu} P = RP$$

$$P = \text{Received light power} \tag{3}$$

In order to achieve a high quantum efficiency $\eta$ the thickness w of the depletion region must be greater than the depth of penetration $x_0$ of the photons. Due to the finite drift times of the generated carriers, however, large layer thicknesses produce a large response time and therefore a small bandwidth with the result that a compromise between sensitivity and bandwidth have to be found in the design of photodiodes (see Section III.A).

## B. Avalanche Photodiode (APD)

In APDs (Figure 4) an additional p region is located between the i region and the n region, where a high electrical field exists ($\geq 10^4$ V/cm). In this p region the charge carriers are accelerated to such an extent that the primary charge carriers generate secondary charge carriers by means of impact ionization. This phenomenon is called the avalanche effect. It leads to an internal current gain $M_0$, which is defined in the following way:

$$M_0 = \frac{\text{total number of charge carrier pairs}}{\text{number of primary charge carrier pairs}}$$

The gain of the APDs depends on the following properties:

- Thickness of the avalanche region
- Relationship between the ionization coefficients of the electrons ($\alpha$) and the holes ($\beta$)

The ionization coefficient is the probability that a charge carrier will produce a new pair of charge carriers per unit distance as a result of impact ionization.[1,2,11] Accordingly, $k_i$ is defined for differing charge carrier injections into the avalanche region:

$$k_i = \beta/\alpha \text{ in the case of electron injection, e.g., Si}$$
$$k_i = \alpha/\beta \text{ with hole injection, e. g., Ge}$$

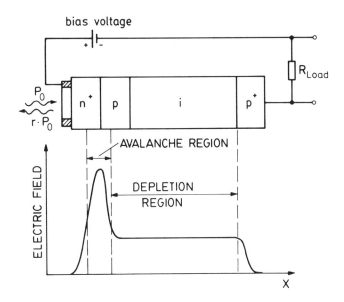

**FIGURE 4.** Structure of an APD and the electric field distribution in the avalanche and depletion region.

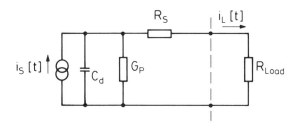

**FIGURE 5.** Equivalent circuit of a photodiode.

As will be shown in the following sections, the $k_i$ values should be small to obtain high bandwidth and low excess noise of the APD. Si for instance, exhibits favorably low $k_i$ values between 0.005 and 0.1, while $k_i$ for Ge is less favorable with values between 0.5 and 1.

Whether the electrons or the holes act as the major ionizing charge carriers depends on the material system used, the crystal orientation, and the electric field distribution in the avalanche region. In the case of ternary and quaternary materials with InP as basic material, either electrons or holes act as the major ionizing charge carriers, depending on the composition of the alloy. In Reference 22 it is shown that the ratio of $\alpha$ to $\beta$ varies between 0.4 for InP and 2 for InGaAs. In case of $Ga_{0.33} In_{0.67} As_{0.7} P_{0.3}$ the ratio of $\alpha$ to $\beta$ is approximately 1.5. Turning to another material system consisting of AlGaAs/GaAs, $\alpha$ and $\beta$ are the same in the basic bulk materials (GaAs $\alpha \simeq \beta$). In Reference 26 a method to enhance the ionization rates ratio by a 50-layer superlattice structure is described, resulting in an $\alpha/\beta$ value of 10.

## C. Equivalent Circuit Diagram

Knowledge of the electrical equivalent circuit diagram of the photodiodes, as shown in Figure 5 is essential in planning the electronic circuitry and in carrying out computer-aided design of the optical receiver (Chapter 8). In this respect there is no basic distinction between PIN and APDs. For most practical cases, the series resistance $R_s$ and the parallel conductance $G_p$ can be ignored.

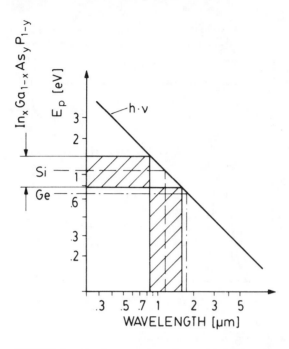

FIGURE 6.   Band gaps and maximum detectable wavelengths for Si, Ge, and the quaternary $In_xGa_{1-x}As_yP_{1-y}$ compound.

The input current pulse $(i_s[t] = e \cdot \delta[t])$ of the photodiodes produces the following current in the load resistor $R_L$:

$$i_L(t) = \frac{e}{\tau} \exp\left(-\frac{t}{\tau}\right) \text{ with } \tau = \frac{R_L + R_S}{1 + G_P(R_L + R_S)} C_d \simeq C_d \cdot R_L \qquad (4)$$

If a response time $\tau \leqslant 50$ psec is to be achieved for a BB transmission system, given a load resistor $R_L$ of 50 $\Omega$, the depletion layer capacitance $C_d$ should attain a maximum of less than 1 pF.

### D. Photodiode Materials for the Optical Long Wavelength Range

Figure 6 shows the maximum wavelengths for which certain semiconductor materials, due to their band gap $E_g$, can be used in photodetectors. For the optical long wavelength range under consideration here, e.g., Si must be ruled out, because of its high band gap of 1.11 eV.

Most of the APDs commercially available today for the wavelengths from 1.3 to 1.6 μm are manufactured using Ge as base material.[7,12] PIN photodiodes made of this material are also available.

The material system composed of $In_xGa_{1-x}As_yP_{1-y}/InP$ is currently the object of intensive investigations.[13-16,22-25] In this system, layers of the quaternary III-V compound are grown, lattice matched on InP substrates. This enables band gaps of between 1.35 and 0.75 eV, in accordance with a wavelength range of $\lambda = 0.9$ to 1.6 μm. PIN photodiodes based on this material system have been commercially available for some time, whereas APD only exist in the form of laboratory samples (see Section III.B.2).

GaAlAsSb[8,9] and HgCdTe[21] are currently being examined regarding their possible use as photodiodes in the long wavelength range.

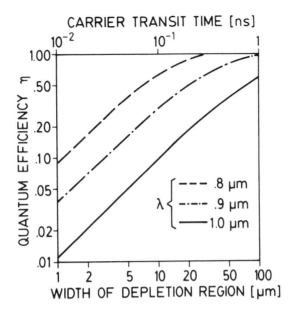

FIGURE 7.    Quantum efficiency as a function of the depletion region width. Parameter is the light wavelength.

## III. PROPERTIES OF PHOTODIODES IN BB TRANSMISSION LINKS

### A. Bandwidths of Photodiodes

The photodiode bandwidth is a very important factor when planning high-rate optical transmission systems. Restrictions on the bandwidth of the detectable base band signals are composed as a result of the transit time of the charge carriers (Section II.A), the capacity of the depletion region (Section II.C), and the frequency-dependence of the multiplication factor M of APDs. Using Si as an example, Figure 7 depicts the quantum efficiency $\eta$ as a function of the depletion region width and the carrier transit time for a PIN photodiode. It is apparent that the achievable quantum efficiency is low in the case of very large bandwidths. On the other hand, with a very small depletion layer width we may expect a high depletion layer capacity, with a result that a compromise between the various parameters must be found when designing a photodiode.

The frequency-dependence of the APD multiplication factor M is described by the following equation:[1]

$$M(\omega) = \frac{M_0}{(1 + M_0^2 \omega_1^2 \tau_1^2)^{1/2}}$$

$M_0$ = DC current multiplication factor

$\tau_1$ = Effective transit time through the avalanche region          (5)

The effective transit time $\tau_1$ again depends on the real transit time $\tau_2$, the ratio of the ionization coefficients $k_i$, and a factor N which varies between 0.33 and 2 depending on the type of material and structure:

$$\tau_1 = N \cdot k_i \cdot \tau_2 \tag{6}$$

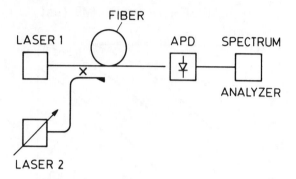

FIGURE 8.    Heterodyne measuring set-up for determining the bandwidth of photodiodes.

For practical applications, the bandwidth or the true frequency response of the photodiodes must be determined by measurements.

However, when determining the bandwidth experimentally, practical problems often occur. The bandwidth can be determined in the time domain or frequency domain, respectively. In the first case, sources emitting very short light pulses are required (pulse width very short compared to the inverse bandwidth of the investigated photodiode).

To measure the frequency response directly, sinusoidal modulated light is used. The modulation frequency is swept from zero to several gigahertz. To be able to reliably determine the frequency response of the photodiode from the received signal, it is essential to have accurate details about the optical signal, i.e., it is either essential to be well acquainted with the modulation properties of the laser used as transmitter, or alternatively, a comparative test must be carried out, using a photodiode with a known frequency response.

In order to measure the bandwidth of photodetectors independently of the modulation properties of the optical transmitter, a heterodyne process is used.[6] Figure 8 shows the basic experimental arrangement. The light of two single-mode (SM) lasers is combined in a directional coupler. Here the beat note of the components with identical polarization is produced and this note is received as an intermediate frequency by the photodiode. If one laser is detuned in comparison with the other, the intermediate frequency can be varied without significantly altering the optical output power. In this way an optical sweep signal can be obtained with sufficiently constant amplitude. The intermediate frequency amplitude is determined by the frequency response of the photodiode and can be measured directly using a spectrum analyzer.

Figure 9 shows the frequency responses of three APDs determined in this way:

- AEG-Telefunken® BPW 28 (0.8 $\mu$m)
- Fujitsu® FPD 150 M (1.3 $\mu$m)
- Alcatel® CG4100 (1.3 $\mu$m)

The results of measurements show that the photodiodes referred to here are not the bandwidth limiting elements in optical Gb/sec systems. The 3-dB bandwidths are in the order of magnitude of 2 GHz.

## B. Noise Considerations in APDs
### 1. Noise Sources

The generation of electron hole-pairs by the absorption of photons and the avalanche gain are statistic processes. This is apparent in the noise behavior of the diodes; the photocurrent and dark current, as well as the avalanche process, are sources of

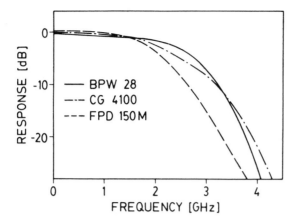

FIGURE 9.    Measured bandwidths of three photodiodes.
Measuring set-up (see Figure 8).

noise. The mean square value of the noise current is described using the following relationship:[10]

$$\overline{i^2} = 2e\,(I_{ds} + [I_{po}F_p(M) + I_{do}F_d(M)]\cdot M^2)B + \overline{i_{na}^2}$$

$I_{ds}$ = Surface dark current outside the avalanche region

$I_{po}$ = Photocurrent at M = 1

$I_{do}$ = Dark current at M = 1

$F_p(M)$ = Excess noise factor of photocurrent

$F_d(M)$ = Excess noise factor of dark current

B = Bandwidth

$\overline{i_{na}^2}$ = Mean square value of the noise current of    (7)

the internal impedance and of the output circuit

The mean square of the photodiode noise current is thus a function of the photocurrent $I_{po}$ and of the dark current components $I_{do}$ and $I_{ds}$. The index "o" describes the so-called primary currents at an avalanche multiplication factor of 1. These primary currents $I_{do}$ and $I_{po}$ pass the avalanche region and form additional charge carrier pairs by impact ionization. By means of this statistic gain process the mean square value of the noise current increases to a greater extent than $M^2$. For this reason excess noise factors are introduced which are different for the photocurrent and dark current noise.

*a. Photocurrent Noise*

Subject to the condition that only one type of charge carrier is injected into the avalanche region, the photocurrent excess noise factor is describd by the following expression:[11]

$$F_p(M) = M \left(1 - (1 - k_i)\left[\frac{M-1}{M}\right]^2\right)$$    (8)

In practice often the rather imprecise approximation is used:

$$F_p(M) \approx M^x \qquad \text{with } 0 < x \leq 1$$    (9)

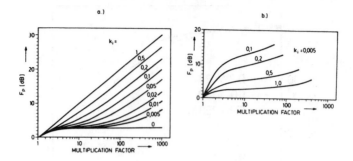

FIGURE 10.    Excess noise factors of the photocurrent as a function of the avalanche multiplication factor.[10] (a) With the injection of the charge carrier type with the higher ionization coefficient. Parameter: ratio of the ionization coefficients; (b) in the case of mixed charge injection. Parameter: ratio of the number of higher ionizing charge carriers to the total number of all charge carriers injected into the avalanche region. The ratio of the ionization coefficients is $k_i = 0.005$.

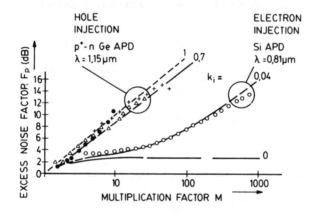

FIGURE 11.    Excess noise factor $F_p$ of two APDs made of Si and Ge, as a function of the multiplication factor M. The excess noise factor of the Ge-APD was determined at 3 different optical levels, producing primary photocurrents of 0.2 $\mu$A (+), 2 $\mu$A (Δ) and 20 $\mu$A (•), respectively.

The above expression for $F_p$ is displayed in Figure 10a for various values of $k_i$, the ratio of the ionization coefficients. The excess noise of an APD declines with decreasing values of $k_i$. For this reason, in the case of Si with its low $k_i$ values between 0.005 and 0.1 compared to Ge with $k_i$ of between 0.5 and 1, very low noise APDs can be produced. Figure 11 illustrates this using typical data for two different APDs. Due to the pronounced spread of the measurement points in case of the Ge-APD, the ratio of the ionization coefficients can only be stated very imprecisely. Using the above mentioned Equation 8 the $k_i$ values are between 0.7 and 1. Thus the photocurrent noise power will increase by approximately the third power of M, if the avalanche gain is increased. This behavior can also be observed in Figure 12 which is described in detail in Section III.B.2. In case of the Si-APD the measured data for $F_p$ results according to Equation 8 in a $k_i$ value of 0.04[18] which is typical for Si. This example illustrates that a low excess noise factor is obtained with those diodes with highly unequal ionization coefficients when only the carrier type with the higher ionization coefficient is injected into the avalanche region (see Figure 10a).

In Reference 10 the relationship for $F_p$ is extended to include the case of mixed injection of charge carriers. Figure 10b, in accordance with Reference 10, shows an

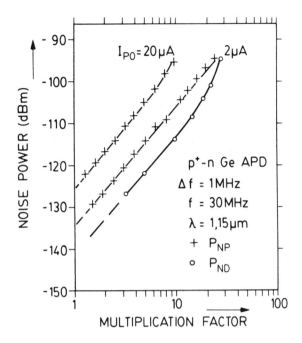

FIGURE 12. Photocurrent $P_{NP}$ and dark current noise $P_{ND}$ as a function of the avalanche multiplication factor M. The photocurrent noise has been measured at two different optical powers, which produced photocurrents of 2 and 20 $\mu A$, respectively, at M = 1. Center frequency f = 30 MHz; bandwidth $\Delta f$ = 1 MHz.

example to illustrate the case of an APD, initially with only one charge carrier type injection (parameter: 1.0).

The excess noise factor increases when the number of charge carriers with high ionization coefficients normalized to all charge carriers decreases because charge carriers with low values of the ionization coefficient are injected simultaneously into the avalanche region.

As mentioned above, in addition to Ge, the InGaAsP, GaAlAsSb, InGaAs, and HgCdTe quaternary and ternary material systems also come into consideration for APDs for the optical wavelength range between 1.3 and 1.6 $\mu m$. In the case of quaternary materials, the band gap of the material can be matched to the photon energy without leading to lattice mismatch.[25] This has a favorable influence on the noise behavior.

APDs of these materials exist up to now only as laboratory samples. Those made of InGaAsP/InP or of InGaAs/InP exhibit excess noise factors of approximately 7 dB at a gain of M = 10.[12,23,24]

APDs made of HgCdTe offer somewhat better noise performance. In accordance with Reference 21 at M = 10 an excess noise factor of 5 dB was achieved.

### b. Dark Current Noise

The dark current causes an additional noise component in the photodiode. The dark current comprises leakage, tunnel, diffusion, and generation-recombination currents. In the case of Si-APDs the dark current is in the pA range[27] (at M = 1) and can therefore be ignored in optical transmission systems. In Ge-APDs dark currents of some microampere are typical, but by reducing the size of the sensitive area 5 nA were achieved.[12] Turning to the ternary and quaternary alloy systems for the wavelength range between 1.3 and 1.6 $\mu$m, the band gap of the material required to achieve high absorption must be as low as possible, whereas for low dark currents it must be selected as high as possible. Current APDs have dark currents in the nanoampere range.

Methods for reducing the dark current consist of using high purity materials in the manufacture of the APD. A reduction of dark current is also achieved in SAM (separated absorption and multiplication)-structures. The APD comprises a narrow band gap region for the light absorption and a separated wide band gap region for the carrier multiplication.[12] The dark current is severely affected by temperature; therefore it is possible to increase the sensitivity of an optoelectronic receiver using simple "Peltier effect cooling".[7,17]

## 2. Noise Properties of Ge-APDs

Ge is used nowadays for many transmission systems in the optical long wavelength range. With test set-ups as described,[2,17,19] the noise components $P_{NP}$ and $P_{ND}$ produced by the photo- and dark currents are determined at a frequency of 30 MHz (Figure 12). The photocurrent noise was measured at two different optical powers, which produced a photocurrent of 2 and 20 $\mu$A, respectively, at M = 1. As can be expected from Figure 11, in both cases the photocurrent noise produced from the signal current is proportional to $M^3$.

The noise power $P_{ND}$ produced by the dark current also behaves in the same way, according to Figure 12, for multiplication factors of <15. With avalanche multiplication factors >15 the dark current noise increases to a much greater extent than with $M^3$. This is due to saturation effects, with the result that different excess noise factors for photo- and dark current must be applied in the expression for the mean square value of the noise current (Equation 7). For high bit rate systems the optimum avalanche multiplication factor $M_{opt}$ is low ($M_{opt} < 15$). In Figure 29 Chapter 8, the signal power $P_s$, the signal-dependent noise power $P_{NS}$, and the dark current noise power $P_{ND}$ of a Ge-APD as well as the amplifier noise power $P_{NA}$ are shown as a function of M. In this graph $M_{opt}$ can be determined at a maximum value of the signal-to-noise ratio (SNR). For increasing bit rates the bandwidth of the system increases, and accordingly the characteristics for $P_{NS}$, $P_{ND}$, and $P_{NA}$ increase. Consequently, to maintain a constant SNR (corresponding to a bit error rate [BER] = $10^{-9}$) the signal power $P_s$ has to be increased, but with increasing signal power $P_{NS}$ increases so that $M_{opt}$ is shifted to lower M-values.

## REFERENCES

1. Melchior, H., Demodulation and photo detections techniques, in *Laser Handbook,* Arrechi, F. T. and Schulz-Dubois, F. D., Eds., Elsevier, Amsterdam, 1972, chap. C7.
2. Stillmann, G. E., Cook, L. W., Bulman, G. E., Tabatabaie, N., Chin, R., and Dapkus, T. D., Long-wavelength (1.3- to 1.6 $\mu$m) detectors for fiber-optical communications, *IEEE Trans. Electron Devices,* 29(9), 1355, 1982.
3. Maslowski, S., New kind of detector for use in communication systems with glass fiber wave guides, Paper WB6, Topical Meet. Integrated Optics — Guided Waves, Materials, and Devices, Las Vegas, 1972.
4. Miller, S. E., Marcatili, E. A. J., and Li, T., Research toward optical-fiber transmission system, *Proc. IEEE,* 612, 1703, 1973.
5. Keiser, G., *Optical Fiber Communications,* McGraw-Hill, New York, 1983.
6. Piccari, L. and Spano, T., New method for measuring ultrawide frequency response of optical detectors, *Electron. Lett.,* 18, 116, 1982.
7. Walker, S. P. and Blank, L. C., Long-wavelength transimpedance optical receiver performance enhancement using cooled germanium avalanche photodiodes, *Electron. Lett.,* 20, 16, 1984.

8. Hildebrand, O., Kuebart, W., Benz, K. W., and Pilkuhn, M. H., $Ga_{1-x}Al_xSb$ avalanche photodiodes: resonant impact ionization with very high ratio of ionization coefficients, *IEEE J. Quantum Electron.*, 17, 284, 1981.

9. Capasso, F., Panish, M. B., and Sumski, S., The liquid-phase epitaxial growth of low net donnor concentration ($5 \times 10^{14} - 5 \times 10^{15}/cm^3$) Ga Sb for detector applications in the 1.3—1.6 μm region, *IEEE J. Quantum Electron.*, 17, 273, 1981.

10. Webb, P. P., McIntyre, R. J., and Conradi, J., Properties of avalanche photodiodes, *RCA Rev.*, 35, 234, 1974.

11. McIntyre, R. J., Multiplication noise in uniform avalanche diodes, *IEEE Trans. Electron Devices*, 13, 164, 1966.

12. Kaneda, T. and Toyama, Y., Avalanche photodiodes for the 1.3—1.6 μm wave band, *Technical Digest*, 4th Int. Conf. Integrat. Opt. Optical Fiber Commun., Tokyo, 1983, 222.

13. Ando, H., Yamauchi, Y., and Susa, N., Reach-through type planar InGaAs/InP avalanche photodiode fabricated by continuous vapor phase epitaxy, *IEEE J. Quantum Electron.*, 20, 256, 1984.

14. Campbell, J. G., Dentai, W. S., Holden, W. S., and Kasper, B. L., High-speed operation of InP/InGaAsP avalanche photodiodes, *Technical Digest*, Conf. Optical Fiber Commun. 1984 (OFC '84), New Orleans, 1984, WA3.

15. Kaneda, T., Advances in InGaAs avalanche photodiodes, *Technical Digest*, Conf. Optical Fiber Commun. 1984 (OFC'84), New Orleans, 1984, WA1.

16. Trommer, R., InGaAs/InP photodiodes with very low dark current and high multiplication, in *Conf. Proc. 9th Eur. Conf. Optical Commun.*, Melchior, H. and Sollberger, A., Eds., Elsevier, Amsterdam, 1983, 159.

17. Kanbe, H., Grosskopf, G., Mikami, O., and Machida, S., Dark current noise characteristics and their temperature dependence in Ge avalanche photodiodes, *IEEE J. Quantum Electron.*, 17, 1534, 1981.

18. Kanbe, H., Kimura, T., Mizushima, Y., and Kajiyama, K., Silicon avalanche photodiodes with low multiplication noise and high speed response, *IEEE Trans. Electron. Devices*, 28, 1337, 1976.

19. Kanbe, H. and Grosskopf, G., Dark current noise properties of a Ge avalanche photodiode, *Jpn. J. Appl. Phys.*, 19, L767, 1980.

20. Brain, M. C., Noise and responsivity of commercial p⁺n Ge avalanche photodiodes, *Electron. Lett.*, 19, 813, 1983.

21. Pichard, G., Meslage, J., Nguyen Duy, T., and Raymond, F., 1.3 μm CdHgTe avalanche photodiodes for fiber optic applications, in *Conf. Proc. 9th Eur. Conf. Optical Commun.*, Melchior, H. and Sollberger, A., Eds., Elsevier, Amsterdam, 1983, 159.

22. Osaka, F., Mikawa, T., and Kaneda, T., Electron and hole ionization coefficient in (100) oriented $Ga_{0.33}In_{0.67}As_{0.7}P_{0.3}$, *Appl. Phys. Lett.*, 45, 292, 1984.

23. Susa, N., Nakagome, H., Mikami, O., Ando, H., and Kanbe, H., New InGaAs/InP avalanche photodiode structure for the 1—1.6 μm wavelength region, *IEEE J. Quantum Electron.*, 16, 864, 1980.

24. Yasuda, K., Kishi, Y., Shirai, T., Mikawa, T., Yamazaki, S., and Kaneda, T., InP/InGaAs buried-structure avalanche photodiodes, *Electron. Lett.*, 20, 158, 1984.

25. Hurwitz, C. E. and Hsieh, J. J., GaInAsP/InP avalanche photodiodes, *Appl. Phys. Lett.*, 32, 487, 1978.

26. Capasso, F., Solid state photomultipliers and avalanche photodiodes with enhanced ionization rates ratio, *Technical Digest*, 4th Int. Conf. Integrat. Opt. Optical Fiber Commun., Tokyo, 1983, 146.

27. Pearsall, T., Photodetectors for communication by optical fibers, in *Optical Fiber Communications*, Howes, M. J. and Morgan, D. V., Eds., John Wiley & Sons, New York, 1980, chap. 7.

28. Baack, C. and Grosskopf, G., Noise properties of opto-electronic receivers in 50 Ω technique, *J. Optical Commun.*, 2, 26, 1981.

Chapter 7

# ASPECTS OF BROADBAND CIRCUIT DESIGN

## Wolfgang Albrecht, Bernhard Enning, and Godehard Walf

### TABLE OF CONTENTS

I.      Special Factors to be Taken Account of for High-Speed Electronic
        Circuits ................................................................................88
        A.      Demands Placed on the Electronic Circuits of a Transmission
                System for High Bit Rates ........................................................88
        B.      Network Analysis as an Aid for Circuit Design ..........................89
        C.      Overview of Commercially Available Hybrid and Integrated
                Components ........................................................................90

II.     Some Properties of Discrete Passive and Active Components ..................91
        A.      Passive Components ..............................................................91
                1.      Capacitors ................................................................91
                2.      Resistors ..................................................................93
                3.      Inductances ..............................................................95
                4.      Transmission Lines ....................................................95
        B.      Active Components ..............................................................95
                1.      Characterizing Parameters ..........................................95
                2.      Bipolar Transistors ....................................................96
                        a.      Small Signal Behavior ........................................97
                        b.      Large Signal Behavior ........................................99
                3.      Field Effect Transistors (MESFET) ...............................105
                        a.      Small Signal Behavior .......................................106
                        b.      Large Signal Behavior .......................................108
        C.      Simulation and Measurement of Implemented Circuits .................108
                1.      Example of Small Signal Behavior .................................108
                2.      Example of Large Signal Behavior ................................111
        D.      Trends for Future Developments ............................................112
                1.      Bipolar Transistors ....................................................112
                2.      Field Effect Transistors (FETs) .....................................112
                3.      High Electron Mobility Transistors (HEMTs) ...................113
                4.      Further High-Speed Active Components for Use in the
                        Gb/sec Range ..........................................................113

III.    Mounting Technology ......................................................................113
        A.      Requirements on Circuits ......................................................113
        B.      Selection Criteria ................................................................113
                1.      Thick and Thin Film Technologies ...............................113
                2.      Synthetic Substrates ..................................................114
                3.      Circuit Technologies for Laboratory Set-Ups ..................114
        C.      Bonding Techniques ............................................................114

References ........................................................................................116

# I. SPECIAL FACTORS TO BE TAKEN ACCOUNT OF FOR HIGH-SPEED ELECTRONIC CIRCUITS

## A. Demands Placed on the Electronic Circuits of a Transmission System for High Bit Rates

In a digital transmission system distinction must be made between analogue and digitally operating circuits. Thus, for instance, the preamplifier, the main amplifier, and the equalizer are to be considered as analogue circuits. If the maximum possible span length and receiver sensitivity are aimed at, correspondingly exacting demands must be placed on these circuits with respect to linearity, frequency response, phase response, noise, and the input and output reflection factors. Likewise, equally high demands must be placed on the amplitude comparator and the timing regeneration so that we can come as close as possible to circuits with ideal properties. Put in other terms, we can say that, although we are dealing with a digital transmission system, similar requirements as to analogue transmission systems must be applied to a large number of the component assemblies if the maximum possible span length is to be achieved. The rules of microwave technology have to be applied on implementation of circuits for a transmission system operating in the gigabit per second (Gb/sec) range. In addition however, further factors must be taken into account in the circuit design, these ensuing primarily from the required large frequency range.

Figure 1 shows the power density spectra of a random pulse sequence with square pulses for two formats (NRZ, RZ). In addition, the power density spectrum for a random RZ pulse sequence with Gaussian-shaped pulses is shown.

The figure shows that the spectra comprise a large frequency range which extends, e.g., for a 1 Gb/sec RZ signal from DC to several gigahertz depending on the pulse shape. This means the electronic components and circuits must be suitable for this very large frequency range and thus must be of very broadband (BB) type.

Hitherto the component market has only to a limited extent catered for digital and analogue circuits of hybrid or integrated types specifically designed for the Gb/sec range. Thus, such circuits have to be developed with discrete components. Since the properties of available components are more or less frequency-dependent, considerable importance must be attached to the process of selecting and characterizing components. Resort can only be made to a limited extent to the large assortment of microwave components obtainable on the market since as a rule these are intended and optimized for a specific frequency range (in general octave bandwidth) which is too narrow for Gb/sec electronics. Likewise, a correspondingly large amount of attention has to be paid to the design of circuits, in particular with respect to parasitic elements (capacitances and inductances). Due to the high-frequency components in the power density spectrum of a Gb/sec signal, such parasitic elements exert a considerable effect on the characteristics of components. They cannot be neglected as it is admissible in general for transmission systems for low bit rates. In contrast to narrow band microwave technology, parasitic elements can only be incorporated or compensated in the circuit concept in specific cases. In general they contribute to a certain low-pass filter behavior and can support tendencies towards oscillation of circuits. Thus the parasitic elements must be kept as small as possible.

A further factor which has to be taken into consideration involves delay times within the connecting lines and components. The delays encountered in practice are of the order of magnitude of 100 psec to several nanoseconds and thus lie within the range of the bit time T of a Gb/sec signal. Delay times play a decisive role in feedback circuits (e.g., quantized feedback equalization) and in circuits in which there has to exist a precise phase relationship between two signals (e.g., timing regeneration). In the second case the necessary phase relationship can be adjusted by delay time compensation.

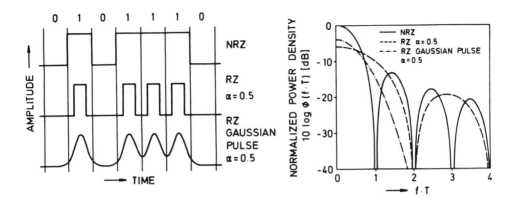

FIGURE 1.    Some pulse forms (left) and power density spectra of random binary sequences (right); $\alpha$ is the duty cycle.

In the case of circuits with feedback, efforts must be made to minimize the delay times. This goal can be achieved by concentrating the design, if possible, by use of integration techniques using very high-speed semiconductors with short delay times, e.g., metal semiconductor field effect transistors (MESFETs) and by minimizing the transistor stages in the feedback circuit.

### B. Network Analysis as an Aid for Circuit Design

As already mentioned in the preceding section, all components possess parasitic elements which at high frequencies cannot be neglected. As a result of this, circuits with only a few components very soon assume the proportions of very complex networks. Moreover rules of thumb for active components as familiar from the field of low-frequency technology lose their validity. In such circumstances a network analysis program can offer major assistance for the design of electronic circuits in the microwave range. By this means the basic behavior of circuits can be determined before they are practically implemented.

The computer-aided design (CAD) program is intended to help simplify circuit analysis for the engineer and relieve him of a large amount of numerical mathematics. In addition this procedure will permit critical features of the circuit concept to be detected so that particular attention can be paid to them during practical implementation of the circuit. As a result of this, valuable microwave components can be protected against damage.

A good CAD program system should permit the following analysis:

1.  DC analysis — DC analysis serves for setting and checking the operating points and any losses which may occur.
2.  DC sensitivity — sensitivity analysis is used to investigate the DC behavior of circuits when components are changed.
3.  DC statistics — the values of manufactured components show a certain spread; the effects of this on the DC behavior of a circuit are calculated by means of DC statistics analysis.
4.  AC analysis — AC analysis is applied to evaluate small signal behavior in the frequency domain.
5.  AC optimization — in the course of AC optimization attempts are made to change the components of a circuit in such a way that a predetermined frequency response is achieved.

## Table 1
### OVERVIEW OF COMMERCIALLY AVAILABLE HYBRID AND INTEGRATED COMPONENTS

Digital Circuits

| Type | Manufacturers | Function | $f_{max}$ | $t_r, t_f$ (20—80%) | $t_p$ |
|------|---------------|----------|-----------|---------------------|-------|
| F100K-serie | Fairchild® | ca. 35 ICs | 300 MHz | 0.85 ns | 0.75 ns |
| 11C06 | Fairchild® | D-flip-flop | 750 MHz | 0.8 ns | 1 ns |
| SP16F60 | Plessey® | OR-gate | 600 MHz | 0.35 ns | 0.55 ns |

Frequency Dividers

| SP8606 | Plessey® | 1:2 | 1,3 GHz | | |
| SP8612 | Plessey® | 1:4 | 2 GHz | | |

Broadband Amplifiers

| UTO1522 | Avantek® | 0.005 — 1,5 GHz | 18 dB | | |
| DC-3002 | B & H | DC — 3,2 Gz | 23 dB | | |
| DC-7000 | B & H | DC — 7 GHz | 21 dB | | |

6.   AC statistics — the effect due to the spread of the values of different components on the frequency response of a circuit is examined with the aid of AC statistics analysis.

7.   Transient analysis — transient analysis is applied to examine circuits within the time domain.

The essential precondition for comprehensive application of an analysis program involves the models set up for both the active and passive components. For the frequency range under consideration, simulation of the components by means of models must be carried out with great care. It requires a great deal of technical measuring equipment and time.

## C. Overview of Commercially Available Hybrid and Integrated Components

In the wake of research and development in the field of military electronics, various manufacturers are now offering analogue BB amplifiers which exhibit a constant frequency response from DC up to the gigahertz range. As an example of the BB amplifiers presently available, mention should be made of the B & H® model DC 7000 H whose specifications extend from DC to 7 GHz with a gain of 21 dB. Pulse rise times of 46 psec can be achieved with this equipment. Other BB amplifier types are listed in Table 1.

Complete digital circuit families can be employed up to a maximum of 300 Mb/sec (Fairchild® F 100k series). Special circuits with individual logic functions such as D flip-flops and OR-gates can be employed for input pulses of up to 1 or 0.55 nsec and maximum frequencies of 750 and 600 MHz, respectively (see Table 1). Frequency dividers for ratios of 1:2 and 1:4 which can be used to 1.3 and 2 GHz, respectively, are offered by Plessey®. In addition it is also possible to have full custom and semicustom designed circuits as bipolar gate arrays and standard cells made to the required specifications of the customer. (For example Fairchild® bipolar digital gate array GE 1000 with toggle frequency of 1 GHz) A good overview of commercially available components is given.[1]

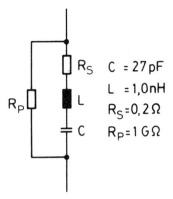

FIGURE 2.    Model of a chip capacitor.

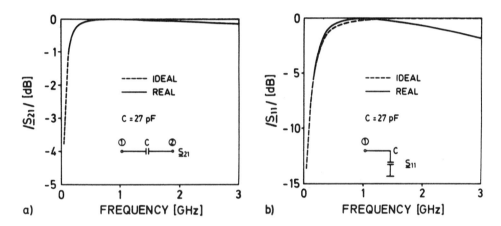

FIGURE 3.    Comparison between real and ideal capacitors. (a) Transmission measurement /$S_{21}$/; (b) reflection measurement /$S_{11}$/. The insets show the measuring method.

## II. SOME PROPERTIES OF DISCRETE PASSIVE AND ACTIVE COMPONENTS

For computer-aided design of active circuits, models have to be developed for both the active and passive components. At frequencies above 1 GHz the passive components employed cease to be ideal. This is taken into account in the discussions in the following sections.

### A. Passive Components
### 1. Capacitors

A model of a multilayer chip capacitor is shown in Figure 2. It consists of an ideal capacitor C, an inductance L in series, and a series resistor $R_s$. The inevitably present leakage is simulated in the model by a resistor in parallel $R_p$. This has the order of magnitude of gigaohms and can be neglected in the following treatment.

The value of C can be clearly determined with an L,C,R measuring bridge. The small values of L and $R_s$ on the other hand cannot be determined by this method since the impedance of the capacity is too high at the measuring frequency (typical 1 MHz).

A further possibility involves determination of these values from the S-parameters measured with the aid of a network analyzer. In the literature, normally transmission and reflexion measurements (insets of Figures 3a and 3b, respectively) are proposed.

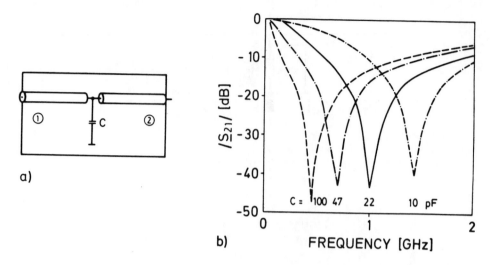

FIGURE 4.    Resonance method for determining the parasitic elements of a real capacitor. (a) Measuring set-up for $\underline{S}_{21}$; (b) measured amplitude responses of $\underline{S}_{21}$ for different capacitors.

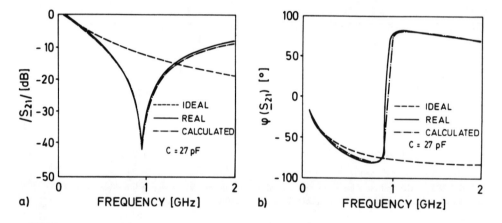

FIGURE 5.    Comparison between real and ideal capacitors for the measuring set-up after Figure 4. (a) Magnitude response of $\underline{S}_{21}$; (b) phase response of $\underline{S}_{21}$.

With the small values of C and $R_s$, hardly any differences can be detected compared with the ideal components by either measurements procedure (Figure 3a, 3b).

These difficulties can be averted by use of the following resonance method. The middle of an ideal 50 Ω microstrip line is connected to ground via the device under test and $\underline{S}_{21}$ of this set-up is measured. The results of measurements taken for various capacity values (typical dimensions 2 × 1 × 1 mm³) are plotted in Figure 4. Pronounced resonance phenomena occur in the relevant frequency range up to 2 GHz. Since C is known, L can be determined from the resonance frequency. The resistance $R_s$ determines the shape and the depth of the resonance curve. Beyond the resonance frequency the component not longer operates as a capacitor but as an inductance. In Figure 5 the amplitude and phase responses of $\underline{S}_{21}$ of the test set-up are represented for an ideal capacitor for the measured components as well as the components calculated in accordance with the model. The deviations between the real and ideal components can be clearly seen in the vicinity of the point of resonance. There is a good agreement between the calculated and measured values. The model thus exactly reproduces the behavior of a real chip capacitor.

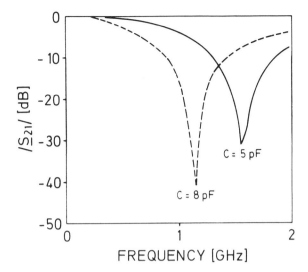

FIGURE 6. Magnitude response of $\underline{S}_{21}$ of a tunable capacitor ("Giga-Trim").

Figure 6 shows the curve as measured for a variable capacitor, i.e., what is termed a "giga-trim". With such components the inductive component is much larger than with the corresponding chip capacitors. Although intended for a frequency range above 1 GHz, only in the case of a tunable capacitor being necessary should this component be employed.

Similar considerations also apply to single-layer GaAs chip capacitors. They can be treated as ideal when no additional lead is needed (e.g., as a coupling capacitor in a coaxial transmission line). In other configurations the required bonding wires will worsen the RF behavior.

In passing it should be mentioned that circuits with different capacitors connected in parallel, e.g., to block high- and low-frequency signal components must be used with caution. In Figure 7 the dashed line shows the behavior of two ideal capacitors connected in parallel whereas the solid line shows the corresponding behavior for real components. Due to the parallel resonance arising, no blocking is encountered at specific frequencies.

## 2. Resistors

For measuring resistors the test method described in the preceding section is slightly modified. The measuring set-up is shown in Figure 8a. The only purpose of $C_M$ is to make the resonance of the test curve more pronounced. The equivalent circuit for a 10 Ω chip resistor (typical dimensions: $0.5 \times 1 \times 2$ mm³) is represented in Figure 8b.

In Figure 9 the amplitude and phase responses of $\underline{S}_{21}$ of the test set-up are shown for an ideal resistor, for the measured resistor, and a resistor calculated in accordance with the model. In this case too, the behavior of a real resistor is reproduced very closely by the model.

The values of the parasitic elements, in contrast to the case with chip capacitors, are relatively small so that their effect in circuits is not particularly significant. Chip resistors can thus be considered as almost ideal components.

The picture is however different when carbon-film resistors are used. A relatively complicated equivalent circuit (see Figure 10a) arises. This component should thus not be used as a resistor at frequencies in the gigahertz range. We have however used it in

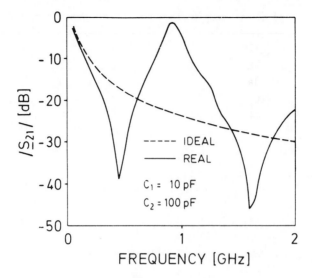

FIGURE 7.     Magnitude response of $\underline{S}_{21}$ for two parallel capacitors.

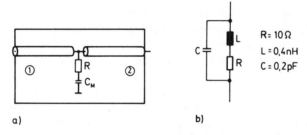

FIGURE 8.     Resonance method for determining the parasitic elements of a chip resistor. (a) Measuring set-up for $\underline{S}_{21}$; (b) model of a chip resistor.

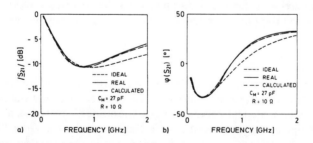

FIGURE 9.     Comparison between real and ideal resistors measured with the set-up after Figure 8. (a) Magnitude response; (b) phase response.

circuits for frequency compensation. The model was derived from direct $\underline{S}_{11}$ measurements with the network analyzer.

Figure 10b shows the calculated and measured curves for a 50-$\Omega$ carbon layer resistor (1/8 W). Considerable deviations from an ideal 50-$\Omega$ resistor are observed with increasing frequencies.

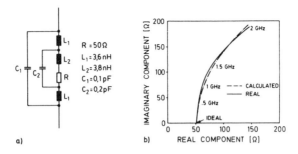

FIGURE 10.    Characteristics of a carbon film resistor (leads
≈ 1 cm). (a) Model; (b) locus of the impedance.

### 3. Inductances

With BB circuits for Gb/sec transmission, consideration should also be given to the possibilities of practical implementation of the design by thin or thick film technology. Since with these technologies it is very difficult if not impossible to produce inductances of more than a few nanohenry, they should not be utilized in the circuit design.

On the other hand, every wire acts as an inductance. For this reason all bonding and connection wires should be kept as short as possible; e.g., a bonding wire with a diameter of 0.025 mm has an inductance of 0.8 nH/mm.

### 4. Transmission Lines

Compared with all the previously described passive components for use in Gb/sec circuits, coaxial cables and planar transmission lines in layered circuits exhibit almost ideal behavior when lengths are restricted to a few centimeters. In Chapter 7, Section II.C and Chapter 9, Sections II and III examples are given where the transformation and delay properties of transmission lines are utilized.[2]

Transmission lines can also be used as quasi-lumped, frequency-independent reactive components if their dimensions are kept small compared with the wavelengths to be processed ($l/\lambda \leqslant 1/8$). They then primarily exhibit:

- Inductive behavior if they, in comparison to their characteristic line impedance, are terminated with lower impedance at both ends
- Capacitive behavior when they, in comparison to their characteristic line impedance $Z_L$, are terminated with higher impedances at both ends

$\pi$ and T-type equivalent circuits for lines with TEM structures are shown in Figure 11. $Z_L$ and v are the related characteristic line impedance and phase velocity. In the case of a coaxial cable $Z_L$ is dependent on the cross-section geometry, and neglecting the relative permeability of the medium between the lines, on its dielectric constant. The phase velocity results solely, again neglecting the relative permeability, from the dielectric constant. In the case of microstrip lines, effective dielectric constants and characteristic line impedances both depend on the geometrical dimensions of the conductor structure and the dielectric constants of the medium between the conductors. Up to several gigahertz the electrical properties can be considered as constant; at higher frequencies they depend on the frequency. Concerning microwave filters, they can act as capacitive and inductive lumped reactive elements.[3] Another example for application is the small feedback capacitor in a BB amplifier.

### B. Active Components
### 1. Characterizing Parameters

If data sheets as issued by transistor manufacturers are to be employed for dimen-

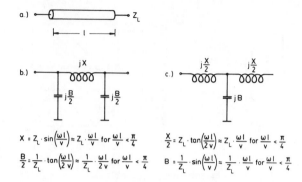

FIGURE 11.    Short transmission line equivalent circuits. (a) Line with TEM structures; (b) π-type equivalent circuit; (c) T-type equivalent circuit; l: line length, v: phase velocity.

sioning BB circuits for use in the Gb/sec range, various difficulties are encountered. Apart from the DC parameters, the short circuit current gain $h_{FE}$ and the transconductance $g_m$ are given. These however, only have limited validity for the behavior in BB circuits. The parameters given for describing the high-frequency characteristics of transistors are the transit frequency $f_T$, the maximum oscillation frequency $f_{max}$, and the stray capacitances, e.g., between collector and base $C_{CB}$ and between emitter and base $C_{EB}$ of a bipolar transistor.

From our own experience, which also agrees with information provided by amplifier manufacturers,[4] the transit frequency does not constitute a suitable selection criterion for developing BB circuits. It may possibly provide information on the time delay in a transistor.[5] The maximum oscillation frequency $f_{max}$, and the stray capacitances $C_{CB}$ and $C_{EB}$ likewise fail to provide a comprehensive description of the behavior of a BB circuit. Only the S-parameters which fully describe the transistor behavior for small signal operation can be drawn upon for circuit analysis and optimization during linear operation. Caution should however also be exercised if S-parameters are taken from data sheets. The S-parameters as determined are heavily dependent on the measuring set-up[6] and may cause the user to arrive at incorrect interpretations of his test results.

If the circuit is driven by large signals there exist no data equivalent to the S-parameters. Only for a few transistor types are data sheets with rise and fall times available.

Attention will here be given to the definition of switching times on the basis of Figure 12. Dependent on the magnitude of the input, signal changes in the delay time $t_d$, the storage time $t_s$, the rise time $t_r$, and the fall time $t_f$ occur; $t_d$, $t_r$, and $t_f$ are defined here by the 10 and 90% values of the output current.

The turn-on time is defined by

$$t_{on} = t_d + t_r,$$

the turn-off time by

$$t_{off} = t_s + t_f$$

## 2. Bipolar Transistors

In general, bipolar transistors were employed for our circuits. In the following an account will be given of the method of deriving transistor models.

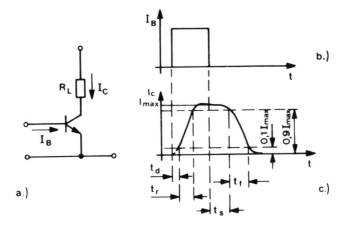

FIGURE 12. Definition of switching times of a transistor. (a) Circuit; (b) input signal; (c) output signal.

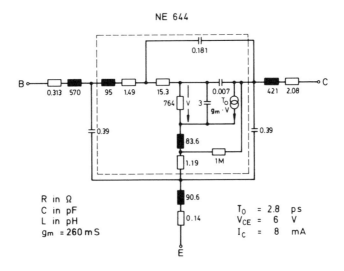

FIGURE 13. Small signal equivalent circuit of the bipolar transistor NE 644.

### a. Small Signal Behavior

For the purpose of modeling transistors, the four S-parameters were measured with the network analyzer for a number of frequencies. An electrical network was derived to fit the curves of the measured $\underline{S}$-parameters. A model was selected which corresponds to the given physical structure of the transistor. The values for the components of the chosen network are determined by means of an optimization program. Figure 13 shows a network consisting of 17 passive and 1 active component. The latter is a current source ($\underline{I} = g_m \cdot V \cdot e^{-j\omega T_0}$). The figure shows the model of the inner transistor (inside the dashed line) and the package.

In Figure 14 the measured $\underline{S}_{11}$ and $\underline{S}_{21}$ parameters are compared with those calculated from the transistor model (NE 644). The agreement is very good.

To serve as examples, the input and output impedances are plotted against the frequency in Figure 15 ($R_G$, $R_L = 50\ \Omega$). Both possess capacitive, reactive components.

Figure 16 shows the calculated short current gain/$\underline{h}_{FE}$/plotted against the frequency.

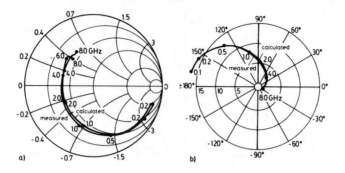

FIGURE 14.    Comparison between measured and simulated S-parameters. (a) Reflection coefficient $\underline{S}_{11}$; (b) transmission coefficient $\underline{S}_{21}$.

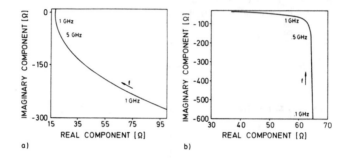

FIGURE 15.    Loci of the input and output impedances of the transistor NE 644. (a) Input impedance; (b) output impedance.

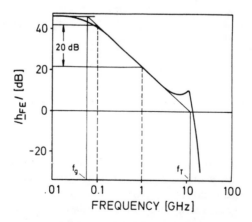

FIGURE 16.    Short current gain $h_{FE}$ of the transistor NE 644.

Above 100 MHz there is a drop of 20 dB/decade which can be observed up to 3 GHz. From this a transit frequency of $f_T = 12$ GHz is arrived at for the NE644. The 3-dB corner frequency $f_g$ is found to be 50 MHz.

Figure 17 shows the calculated pulse response for a square wave input impulse. The calculation was carried out for a transistor in common emitter configuration with $R_L = 50\ \Omega$.

Figure 18 depicts the small signal equivalent noise circuit derived from the equivalent

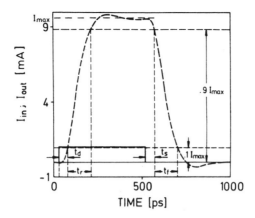

FIGURE 17. Pulse response of the transistor NE 644 in common emitter configuration. Solid line: input signal; dashed line: output signal.

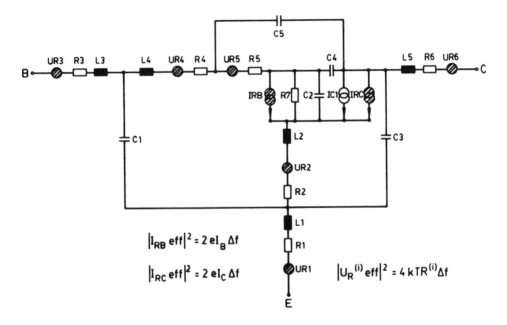

$$|I_{RB}\,eff|^2 = 2\,eI_B\,\Delta f$$

$$|I_{RC}\,eff|^2 = 2\,eI_C\,\Delta f$$

$$|U_R^{(i)}\,eff|^2 = 4\,kTR^{(i)}\Delta f$$

FIGURE 18. Equivalent noise circuit of the bipolar transistor NE 644; noise sources are hatched.

circuit of Figure 13. The thermal noise of the resistors and the Schottky noise are taken into account. Flicker noise which arises at low frequencies is neglected. Furthermore, it is assumed that the sources of noise are uncorrelated with each other. The advantage of this noise model lies in the fact that the sources of noise are located where they are located in reality.

In conclusion, it should be pointed out that these equivalent circuits only apply for the corresponding operating point at which the $\underline{S}$-parameters were measured.

### b. Large Signal Behavior

The equivalent circuit for small signals is not suitable for describing the nonlinear behavior of transistors. For this reason an equivalent circuit for large signals will be developed below.

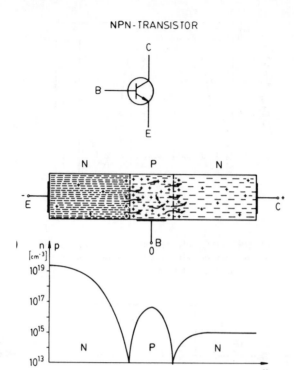

FIGURE 19.    NPN-transistor. (a) symbol; (b) principal structure; (c) doping profile.

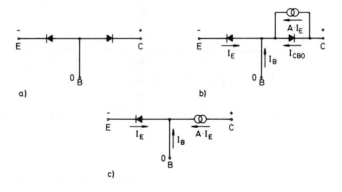

FIGURE 20.    Steps of modeling the transistor for large signal behavior.

On account of its symmetrical structure of layers (NPN) the transistor appears to be
a component with interchangeable emitter and collector connections. However, on account of the differences in doping of the zones (Figure 19) the transistor is unsymmetrical. If the emitter and collector connections are swapped this will lead to major
changes in its characteristics.

The two PN-junctions are initially simulated by two diodes (Figure 20a). The polarity relationships are correctly reproduced but the control of the collector current by the
base current is not taken into account. In spite of the reverse biased base-collector
diode a large minority carrier current is injected from the emitter through the small
base region to the collector. The equivalent circuit must thus be supplemented by a

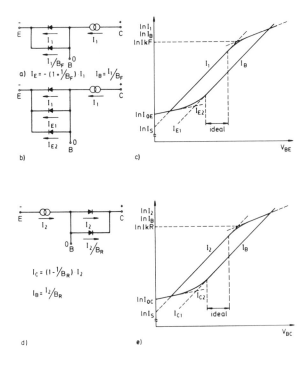

FIGURE 21.    Improved model of the transistor. (a,b) base emitter diode forward biased; (c) $I_1$, $I_B$ vs. $V_{BE}$; (d) base collector diode forward biased; (e) $I_2$, $I_B$ vs. $V_{BC}$.

current source connected in parallel to the collector diode and supplying a current A · $I_E$ ($/A/ \approx 1$) (see Figure 20b).

The current A · $I_E$ is very large compared with $I_{CBO}$ which can thus initially be ignored. The results in the circuit of Figure 20c is then arrived. The current $I_E$ is now split into current $I_1$ and the base current $I_B$ (Figure 21a). In this connection $\beta_F$ is the forward current gain. The current through the upper diode is equal to the current of the current source.

The general relationship between $I_1$ and $I_B$ in dependence of $V_{BE}$ is shown in Figure 21c[7-9] for a real transistor. Only the ideal middle part of the curve is simulated by the model of Figure 21a.

At low voltages the base current deviates from a straight line. This is due to the following three factors:

1.    Recombination of the carriers on the surface
2.    Recombination of the carriers at the emitter base junction
3.    Existence of surface channels from the emitter to the base

These parts of the base current are always present; however, at relatively large values of $V_{BE}$ they can be neglected. The base current is thus composed of $I_{E1} + I_{E2}$ which can be described by two ideal diode characteristics. These are two straight lines with different slopes in a logarithmic scale. In the model the curves are represented by one diode in each case (Figure 21b and c).

If the polarity of emitter and collector is now changed, the base collector junction determines substantially the behavior of the transistor. This PN-junction is modeled in the same manner (Figure 21d and e) as the base emitter junction. The different doping proportions can be taken into account in the equations. The whole transistor can be modeled by superposition of both equivalent circuits (Figure 22).

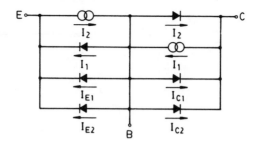

FIGURE 22.    Equivalent circuit for both PN-junctions of the transistor.

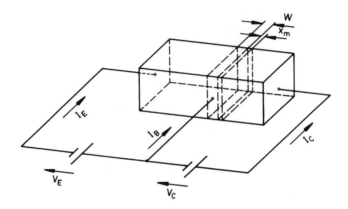

FIGURE 23.    Simplified drawing of the one-dimensional transistor.

The next step is to simulate what is termed the "Early Effect".[10] We consider again an NPN-transistor in the common emitter configuration. The PN base-emitter junction is biased in the forward direction. Therefore the space-charge region between the base and the emitter can be ignored. The base-collector diode is reverse biased, i.e., the barrier thickness of the space charge region $x_m$ is a function of the reverse voltage (Figure 23). Since the space-charge region expands towards each side with increasing reverse voltage, the base width w becomes smaller.

Reducing the base width leads to the following effects:

1.    The recombination of injected minority carriers in the base decreases since the minority carriers on an average diffuse across the narrower base in a shorter time.
2.    The impedance for the minority carrier current injected by the emitter changes. The impedance "seen" by this current depends on the base layer resistivity and its thickness. A reduction in the thickness results in a reduction of the impedance.

The influence of these effects is shown in Figure 24. Note that for an ideal transistor, $I_1$ is independent of $V_{CE}$. The intercept of the dashed straight lines with the $V_{CE}$-axis is the so called Early Voltage $V_A$. The Early effect for normal and reverse operation is taken into account by an additional current source $I_3$ (Figure 25).

The last factor in the equation for $I_3$ takes into account high current effects which will be investigated now. Since the following effects require a very comprehensive mathematical description,[7] only a brief qualitative description will be given here. Figure 21c is considered again. For high voltages $V_{BE}$ the collector current bends away from a straight line. This is due to high current effects which arise from the large injection of electrons from the emitter into the base.

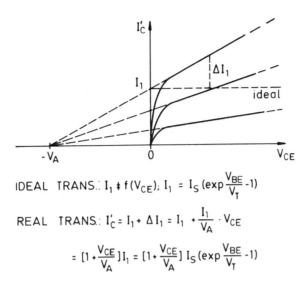

IDEAL TRANS.: $I_1 \neq f(V_{CE})$; $I_1 = I_S (exp\frac{V_{BE}}{V_T} -1)$

REAL TRANS.: $I'_C = I_1 + \Delta I_1 = I_1 + \frac{I_1}{V_A} \cdot V_{CE}$

$= [1 + \frac{V_{CE}}{V_A}]I_1 = [1 + \frac{V_{CE}}{V_A}] I_S (exp\frac{V_{BE}}{V_T} -1)$

FIGURE 24. Influence of the Early effect on the $I_c'$-$V_{CE}$ characteristic.

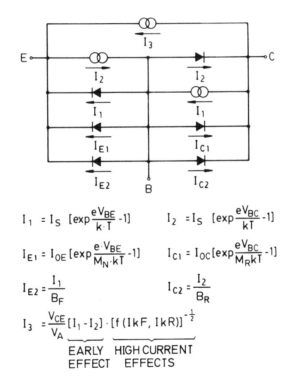

$I_1 = I_S [exp\frac{eV_{BE}}{k \cdot T} -1]$       $I_2 = I_S [exp\frac{eV_{BC}}{kT} -1]$

$I_{E1} = I_{OE} [exp\frac{e \cdot V_{BE}}{M_N \cdot kT} -1]$       $I_{C1} = I_{OC}[exp\frac{eV_{BC}}{M_R kT} -1]$

$I_{E2} = \frac{I_1}{B_F}$       $I_{C2} = \frac{I_2}{B_R}$

$I_3 = \underbrace{\frac{V_{CE}}{V_A}}_{\substack{EARLY \\ EFFECT}}\underbrace{[I_1 - I_2] \cdot [f(IkF, IkR)]^{-\frac{1}{2}}}_{\substack{HIGH CURRENT \\ EFFECTS}}$

FIGURE 25. Modeling of Early and high current effects; currents are explained in Figure 21c. IkF and IkR are the forward and reverse high injection knee currents (see Figure 21).

Up to now the static state of the transistor has been considered. The dynamic behavior of the transistor will now be investigated.

If an AC voltage is applied to a reverse biased PN-junction, the thickness of the space charge region and thus also the magnitude of the space charges changes. This

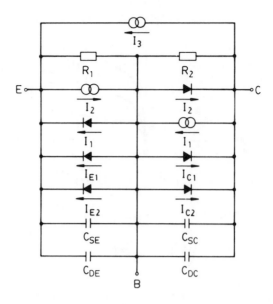

FIGURE 26.    Transistor equivalent circuit including junction capacitances ($C_{SE}$, $C_{SC}$), diffusion capacitances ($C_{DE}$, $C_{DC}$), and leakage current effects ($R_1$, $R_2$).

effect is simulated in the model by a nonlinear junction capacitance (Figure 26) which depends on the physical structure of the PN-junction. There are two limits for the physical structure of a PN-junction: the abrupt junction and the linear junction. Most junctions in practice lie somewhere between these two limits.

The junction capacitance $C_{SE}$ of the reverse biased base emitter diode is[8]

$$C_{SE} = f(V_{BE}, \Phi_E, m_E, C_{SEO})$$

$V_{BE}$ = Emitter-base voltage

$\Phi_E$ = Diffusion voltage, base-emitter

$m_E$ = Gradient of emitter-base capacity

This is a function of the PN-boundary

Abrupt boundary:     $m_E = 0.5$

Linear boundary:     $m_E = 0.333$

$C_{SEO}$ = Value of the base-emitter junction

capacity for $V_{BE} = 0$

Similar considerations apply for the junction capacitance $C_{SC}$ of the base-collector PN-junction. The junction capacities thus describe the effect of the voltages acting across the individual PN-junction.

$$C_{SC} = f(V_{BC}, \Phi_C, m_C, C_{SCO})$$

The next step is to consider the forward biased base emitter diode. In the transition from the reverse to the forward state, the PN-junction is flooded by majority carriers which diffuse into the regions of opposite doping. There they recombine as minority carriers or they reach the collector. Near the PN-junction in the P-region electrons are stored and in the N-region holes are stored. Due to the storing effects, the PN-junction represents a capacity. In the model this is taken into account by the diffusion capacities $C_{DE}$ and $C_{DC}$ (Figure 26). These are nonlinear capacities.

Diffusion capacities:

$$C_{DE} = g(V_T, \tau_F, I_C)$$

$$V_T = \frac{kT}{e} \quad \text{Thermal voltage}$$

$$\tau_F = \text{Forward transit time}$$

$$I_C \simeq \text{Collector current}$$

$$C_{DC} = g(V_T, \tau_R, I_E)$$

$$\tau_R = \text{Reverse transit time}$$

The leakage currents arising with every PN-barrier are simulated by $R_1$ and $R_2$ (Figure 26).

The ohmic path resistances of the transistor must also be taken into account. The internal emitter resistance is simple to simulate in a model (Figure 27). The resistor $R_E$ can be considered as constant for all ranges of operation. The internal collector resistance $R_C$ varies with the operating point.[9] To simulate the internal base resistance is more difficult. As already described with the Early effect it is markedly influenced by the base width modulation. It can be considered as a nonlinear resistor (dashed line) or it can be split into a constant inner $R_{Bi}$ and outer base resistor. But this assumption with two constant components is only an approximation. Between the electrical contacts on the chip, parasitic capacities exist which are inserted into the transitor model as $C_{BE}$, $C_{CE}$, and $C_{BC}$. This completes the model of the transistor chip.

For packaged transistors the influence of the packaging must be taken into account in the model. The inductances of the bonding wires ($L_{BB}$, $L_{CB}$, $L_{EB}$) and leads ($L_{BA}$, $L_{CA}$, $L_{EA}$) as well as the capacities ($C_{BA}$, $C_{CA}$, $C_{EA}$) must be taken into consideration (Figure 28).

The temperature-dependence of the transistor is included explicitly in the equations for the equivalent components.

In conclusion, two restrictions which apply to the model should be mentioned:

1. The breakdown behavior has not been included in the model.
2. Only a few boundary layer and surface effects have been taken into account.

### 3. Field Effect Transistors (FET)

In contrast to bipolar transistors, the properties of a field effect transistor are only determined by charge carriers of one polarity (i.e., electrons or holes). In this chapter only the fastest FET, the metal semiconductor field effect transistor (MESFET), is investigated. With this transistor the input diode between the gate and source is operated in the reverse state. For this reason the input impedance is very large for lower frequencies since only the very small minority carrier reverse current exists. Therefore a negligibly small input power is required to control the MESFET. For low frequencies

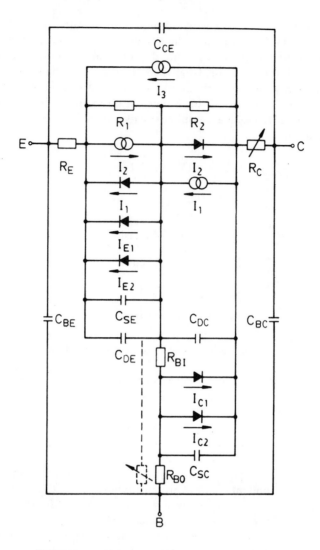

FIGURE 27.    The complete model of the transistor chip.

the output impedance also has high values. The transconductance of a MESFET ($g_m$ = 20 to 40 mS) is much less than in the case of bipolar transistors ($g_m$ = 300 to 500 mS). However, in their noise behavior and speed they are superior to these.

### a. Small Signal Behavior
In this section a physical equivalent circuit for the small signal behavior is presented without further consideration of the derivation (Figure 29). Only the most important sources of noise are considered.

Figure 30a and 30b represent the input and output impedances as a function of the frequency. The input impedance behaves like a series circuit consisting of a resistor and a capacitor. For high frequencies neither the input nor the output impedances have very high values. If a transit frequency is defined corresponding to the bipolar transistor, the transit frequency of the MESFET (MGF 1800) is $f_T$ = 10 GHz (Figure 31).

The decrease of the short current gain $h_{FE}$ again amounts to 20 dB/decade. But this decrease starts at low frequencies like an integrator.

Figure 32 shows the calculated pulse response of the MESFET in common source

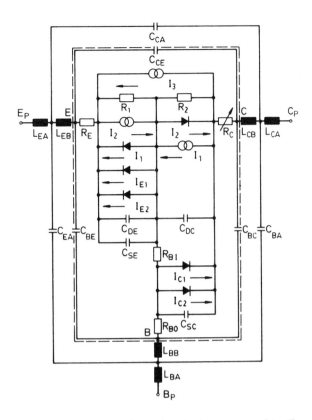

FIGURE 28. Model of a packaged microwave transistor (In dashed line: transistor chip).

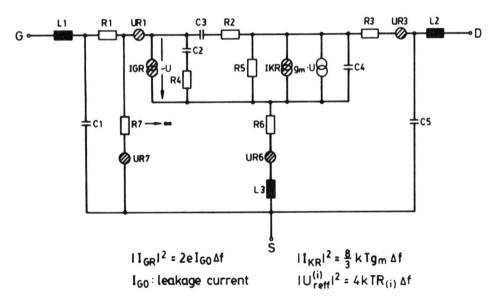

$$|I_{GR}|^2 = 2e I_{G0} \Delta f$$

$I_{G0}$: leakage current

$$|I_{KR}|^2 = \frac{8}{3} k T g_m \Delta f$$

$$|U_{reff}^{(i)}|^2 = 4 k T R_{(i)} \Delta f$$

FIGURE 29. Small signal equivalent circuit of a MESFET with noise sources included.

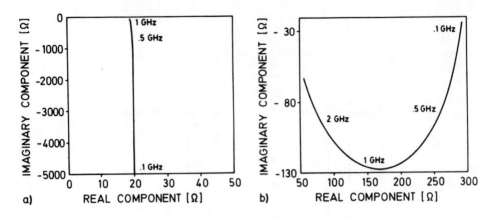

FIGURE 30.    MESFET in common source configuration. (a) Input impedance; (b) output impedance.

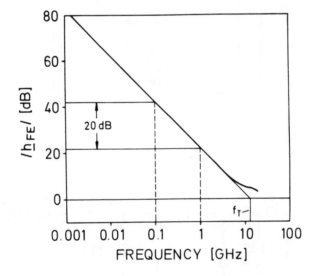

FIGURE 31.    Short current gain $h_{FE}$ of a MESFET vs. frequency.

configuration with $R_L = 50 \, \Omega$. In comparison with Figure 17 it can be seen that the
MESFET is much faster than the bipolar transistor. On the other hand the gain is much
smaller.

### b. Large Signal Behavior

Without further consideration of the derivation an equivalent circuit (Figure 33) is
indicated with which the large signal behavior of a MESFET can be modeled. An N-
channel transistor is assumed. The two diodes simulate the two gate junctions in the
model. The ohmic resistance of the drain and source regions are taken into account by
$R_D$ and $R_S$. The current source $I_{DS}$ simulates the channel current. The charges stored in
the two gate junctions are simulated in the model by the two capacitors $C_{GD}$ and $C_{GS}$.

### C. Simulation and Measurement of Implemented Circuits

### 1. Example of Small Signal Behavior

An amplifier with three current feedback stages with bipolar transistors[12] will be

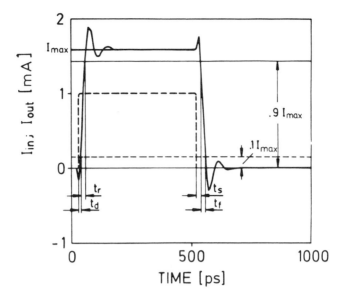

FIGURE 32. Pulse response of the MESFET (MGF 1800) in common source configuration. Dashed line: input signal; solid line: output signal.

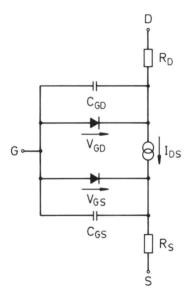

FIGURE 33. Large signal equivalent circuit of a MESFET.

used to demonstrate that all parasitic elements of active and passive components have to be taken into account when analyzing and optimizing BB circuits. Figure 34 shows the basic circuit diagram of the amplifier. This is an example for a main amplifier in a repeater. Its noise characteristics here play a subordinate role and will thus not be considered. Since lossless BB low reflection matching is not possible,[13] the resistive input voltage divider consisting of the 33 and 18 $\Omega$ resistors was provided. The networks in dashed lines are for correcting the frequency response. They consist of chip resistors $R_{1,2,3}$ and open-ended transmission lines $T_{1,2,3}$.

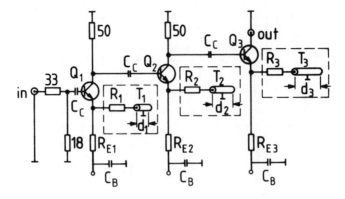

FIGURE 34.    Three-stage BB amplifier.

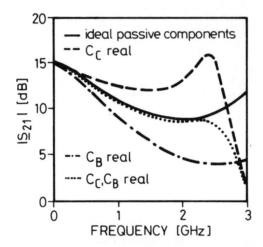

FIGURE 35.    Magnitude of $\underline{S}_{21}$ of the amplifier without frequency response correcting networks.

Omitting the frequency response, correcting networks calculations were undertaken to demonstrate the influence of the real coupling and blocking capacitors $C_C$ and $C_B$ upon the transmission factor $\underline{S}_{21}$ (see Figure 35). Under the assumption of real transistors and ideal passive components the solid line was obtained. The dotted line shows the response calculated with real transistors and real passive components. The agreement between these curves in the range 0 to 2 GHz is accidental since the influence of the parasitic elements of coupling capacitors and blocking capacitors just compensate each other. This is clearly seen from the dashed curves where real coupling capacitors $C_C$ and ideal blocking capacitors $C_B$ were assumed and from the dashed dotted curve where real blocking capacitors $C_B$ and ideal coupling capacitors $C_C$ were chosen.

Figure 36 shows the input stage of the circuit, taking into account all parasitic elements of the passive components. For clearance the transistor equivalent circuit and the other stages are omitted. Values for $C_R$ are 0.1 to 0.3 pF; $L_R$ and $L_C$ are 0.5 to 3 nH; $L_B$ is 9 nH and $R_C$ is 0.2 to 0.4 Ω.

In conclusion, Figure 37 shows the amplitude and phase responses of the whole three-stages amplifier including the optimized frequency response compensating networks. Continuous lines depict the calculated ones with all parasitic elements, the dotted lines the measured ones. Comparison of the curves shows that there is good agreement between the calculations and measurements.

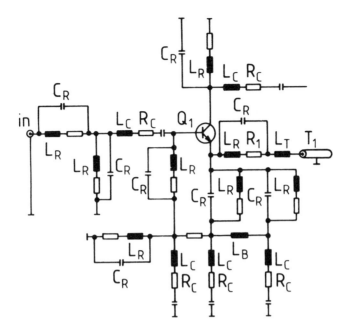

FIGURE 36. Input stage including all parasitic elements of the passive components.

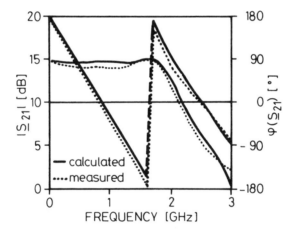

FIGURE 37. Amplitude and phase response of the complete optimized amplifier.

## 2. Example of Large Signal Behavior

The following example gives a comparison between simulation and measurement of a current mode switch with emitter followers at the outputs. The circuit represented in Figure 38a is designed with transistor chips of type BFR92 ($f_r$ = 5 GHz) and chip resistors on a thick film substrate. The four transistors are incorporated in a 16-pole chip carrier.

For measurement and simulation, an input signal with a peak to peak voltage of 250 mV (Figure 38b, dashed line) and a rise time of 600 psec (10 to 90%) was used. The solid curve represents the measured output signal and the dotted curve the calculated one for the right output. The agreement between calculation and measurement is very good.

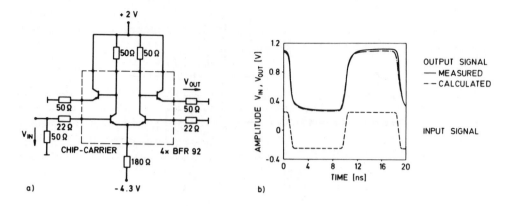

FIGURE 38.    Current mode switch. (a) Circuit configuration; (b) time domain behavior, measured and calculated.

## D. Trends for Future Developments

Due to the increasing digitalization of communication networks, the rapid development of optical communications technology, and the need for high-speed computers, a demand for high-speed components extending beyond the Gb/sec range is anticipated.[14-17] In addition to Si and GaAs semiconductors used as bipolar and field effect transistors (FETs), other semiconductor materials and structures such as InP and high electron mobility transistors (HEMTs) have been developed. They will also influence high-rate optical communications technology. Other high-speed circuit components such as Josephson elements and tunnel diodes which are likewise suitable for digital communication, will however not be used at all for optical communications, because of the enormous cooling requirements for the Josephson elements and the shortcomings of the tunnel diodes, i.e., instabilities and lack of reproducibility.[18]

Concerning integration of semiconductor circuits, semicustomed and full customed design will advance into the Gb/sec range.[19,20]

### 1. Bipolar Transistors

Presently, commercially available circuit families can be employed up to about 300 Mb/sec. They are built up on the basis of bipolar technology with Si transistors (F 100K series). From laboratory results already published it can be assumed that this technology will be used for circuits extending into the Gb/sec range.[21-23] This literature presents word generators for up to 3 Gb/sec, frequency dividers up to 5.5 GHz clock frequency, and multiplexers up to 1 Gb/sec. A bipolar transistor based on GaAlAs-GaAs technology is mentioned which can be implemented for the Gb/sec range.[24] An example of an InP bipolar transistor is given.[25]

### 2. Field Effect Transistors (FETs)

Recently, high-speed integrated circuits with FETs based on GaAs and to some extent on InP technology have appeared in competition to bipolar circuits. The advantages are lower power consumption and higher attainable speeds due to a six fold increase in the mobility of electrons compared with Si[26], and lower noise for use in amplifiers. On account of their possible integration with optoelectronic components, InP and GaAs semiconductors are of considerable significance.[27-30] A large number of publications deal with the implementation of circuits based on GaAs technology for use in the Gb/sec range.[31-39]

In addition to BB amplifiers up to 12 GHz, digital circuits are also presented which can operate up to 5 Gb/sec. To a large extent these involve monolithic microwave integrated circuits (MMICs). The technology of InP transistors can still be considered to be in its infancy. This is evident from the relatively small number of publications.[40] Use is possible up to the microwave range.[27,41]

### 3. High Electron Mobility Transistors (HEMTs)

The fastest transistors at present are HEMTs. The technology is still however, at the start of the laboratory stage. The electron mobility of a GaAs HEMT is about five times faster than that of GaAs MESFETs. The circuits presented include ring oscillators and frequency dividers.[40,42] It can, however, be expected that the HEMTs will be used in the future in high-speed computers and in communications technology as high-speed digital components. On the other hand, for optical communications it is of significance that they can also be realized in InP technology.

### 4. Further High-Speed Active Components for Use in the Gb/sec Range

Brief mention should be made of Josephson elements and step recovery diodes. Whereas Josephson elements are able to perform maximum-speed switching operations (approximately 10 psec)[43], clock frequencies > 10 GHz are propagated[14] — they will hardly be used at all in communications systems since the operational temperatures have to be kept down to a few degrees K. Step recovery diodes have been employed in multiplexers for transmission rates of 16 Gb/sec[44] in hybrid circuits.

## III. MOUNTING TECHNOLOGY

### A. Requirements on Circuits

If 2 Gb/sec signals are to be transmitted, the lower cut-off frequency may lie in the acoustic range, whereas the upper cut-off frequency lies in the microwave range. Therefore the geometrical dimensions of hybrid circuits approach the order of magnitude of the wavelength of the high-frequency spectral signal components. As basic principle, the discrete components have to be built up as close as possible. Parasitic elements due to connecting leads and soldering have to be kept small. For connecting electronic circuits, symmetrical or unsymmetrical conductor structures for wave guidance such as stripline, microstripline, and coaxial cables are required.

It is assumed that the reader is familiar with the dimensioning of planar microwave circuits. However, mention is made of a publication which sums up all the most important equations for dimensioning microstrip lines.[45] Another publication[46] also provides information on the predictability of discontinuities in conductor routes.

### B. Selection Criteria

In the choice of one of these technologies, further factors must be taken into consideration such as the convenience of processing, power dissipation, costs, etc.

For hybrid circuits with integrated and discrete components, unsymmetrical wave guiding structures will be of advantage, since the requirements for sticking, soldering, and bonding are less. Likewise, machining of the carrier material is simplified. MMICs on account of the enormous development costs, are only justified if high production runs can recoup the high development costs.[47,48]

### 1. Thick and Thin Film Technologies

Owing to the technological requirements and the maximal usable frequency, distinction is made between thick and thin film technologies. Whereas thick film circuits are frequently produced by easy-to-apply screen printing methods, the facilities required for thin film technology are in general much more complicated. The metallization must be applied by complicated evaporation methods or cathode sputtering, and the structures are achieved by etching. The maximum usable frequency of thick film circuits is restricted by the roughness of the substrate surface and the inaccuracy of the printed lines. For example, at the conductor edges where the current density is maximum, this leads to considerable attenuation and radiation losses. Thick film circuits are used up to 5 GHz. Thin film circuits can be used for frequencies far above 5 GHz. On account of the geometrical dimensions of thin film circuits, even directional couplers and filters with distances between conductors in the micron range can be produced. A good comparison between thick and thin film techniques is given.[49-51] This literature also provides detailed descriptions of the individual technological processes.

Ceramic materials are generally used as a substrate for thick and thin film technologies whereby higher demands are placed on the surface side for the thin film technology. The normal materials are aluminium oxides $Al_2O_3$ with dielectric constants between 9 and 10. However, their tenacity represents a disadvantage since they can only be processed mechanically with special tools and devices such as diamond drills or gas lasers. On the other hand, their high thermal conductivity constitutes an advantage. In addition, materials consisting of glass, sapphire, and beryllium are used for thin film substrates.

## 2. Synthetic Substrates

If the demands on bending resistance and thermal conductivity are not so stringent, substrates consisting of polyolefins reinforced by fiber glass can be used. In addition, materials made from Teflon® or polystyrol can be employed. In comparison to ceramic materials and mixtures of synthetic materials and ceramics, these possess a lower dielectric constant ($\varepsilon_r = 2$ to $9$). This leads to larger dimensions for the conductors and to spatially extended electromagnetic fields around the conductors. This requires larger distances from neighboring conductors to prevent undesirable coupling. It also means that the maximum possible packing density in such circuits is low.

## 3. Circuit Technologies for Laboratory Set-Ups

In addition to the above-mentioned implementation technologies which are suitable for mass production in industry, there also exists a very flexible system permitting very quick hardware implementation of circuits, e.g., Minimount®. This is a kind of "bread-board" system for microwave circuits. The basic components of Minimount® are a two-sided, metal-plated synthetic board (a) which serves as a chassis (Figure 39) and a synthetic strip (b) coated on one side with metal. The metal-coated synthetic strip and the metallic chassis ground form a waveguide with a defined characteristic impedance. It is secured to the chassis board with the aid of an adhesive. Both can be used as a solder base for semiconductors and passive components. This system has been used successfully for frequencies up to 2 GHz. A practically implemented example of an amplifier is shown in Figure 40A. For comparison, a circuit in thick film technology is given in Figure 40B.

## C. Bonding Techniques

Among other attachment techniques a few remarks shall be made here concerning wire bonding. Further information is recommended.[52]

If nonpackaged semiconductors and integrated components are used, the connections to other components are established by wire bonding techniques. The normal procedure employs wire connections with an inherent high inductance (see Chapter 7, Section II). For connections with a low inductance, ribbons with a mesh structure are employed (mesh bonding). Distinction is made between thermocompression bonding and ultrasonic bonding. The mechanical and electrical connections from the wire to the metal-plated electrodes in the case of both bonding techniques are based on micro-

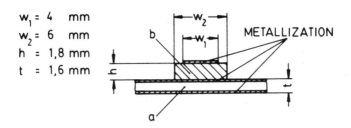

FIGURE 39. Cross-section of a waveguide in Minimount® technology.

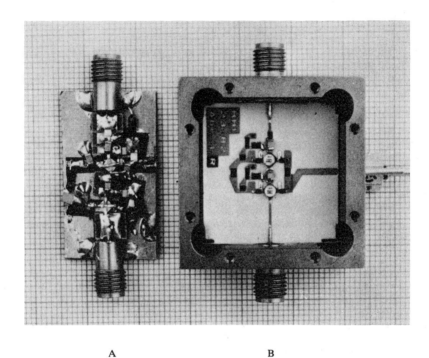

A                       B

FIGURE 40. Photograph of a two-stage amplifier. (A) Minimount® circuit; (B) thickfilm circuit.

scopic fusion on the surface of the two materials to be connected. Due to intensive mechanical contact of the two materials, adhesion forces are produced. In the case of ultrasonic bonding, the ultrasonic movement of the bond wire against the conductor pad rubs off the oxidation layers of the surface. Some metals can be connected together which may be of nonnoble types. Noble metals without oxidized surfaces are more suitable for thermocompression bonding on account of the sensitivity of this method to surface impurities. The choice of method depends on the type of metallization of the semiconductors and passive components electrodes, e.g., hardness and surface roughness. The best bond type for the specific case is normally indicated by the manufacturers of the semiconductor.

# REFERENCES

1. Overview over standard cells and bipolar digital arrays, *EDN,* 28, 104, 1983.
2. Kraus, J. D. and Carver, K. R., *Electromagnetics,* McGraw-Hill, New York, 1973.
3. Mattaei, G. L., Young, L., and Jones, E. M. T., *Microwave Filters, Impedance-Matching Networks and Coupling Structures,* McGraw-Hill, New York, 1964, 355.
4. Avantek, High-frequency transistor primer. I—III. Application note of Aventek, ATP-1018/R-1-82; ATP-1047/R-10-82; ATP-1040/R11-82.
5. Barna, A., Propagation delay in current mode switching circuits, *IEEE J. Solid State Circuits,* 8, 123, 1975.
6. Manz, B., Three-firm collaboration yields new "standard test fixture", *Microwaves,* 21, 19, 31, 1982.
7. Sze, S. M., *Physics of Semiconductor Devices,* John Wiley & Sons, New York, 1969.
8. Getreu, I., *Modelling the Bipolar Transistor,* Tektronix, Beaverton, Oreg., 1976.
9. Philpac Model Book, Philips, Eindhoven, The Netherlands, 1982.
10. Early, J. M., Effects of space-charge layer widening in junction transistors, *Proc. IRE,* 40, 1401, 1952.
11. Nagel, L. W., *Spice 2: A Computer Program to Simulate Semiconductor Circuits,* University of California, Berkeley, 1975.
12. Albrecht, A. and Enning, B., A broadband bipolar transistor amplifier for Gbit/s applications, developed with a CAD optimization, *Frequenz,* 38, 121, 1984.
13. Fano, R. M., Theoretical limitations on the baseband matching of arbitrary impedances, *J. Franklin Inst.,* 249, 57—84, 139—154, 1950.
14. Bosch, B. G., Device and circuit trends in gigabit logic, *Proc. IEEE,* 127, 254, 1980.
15. Greiling, P. T. and Waldner, M., Future applications and limitations of digital GaAs IC technology, *Microwave J.,* 26, 74, 1983.
16. van Thyl, R. and Liechti, C., Gallium arsenide spawns speed, *IEEE Spectrum,* 14, 41, 1977.
17. Research Aspects of the Gallium Arsenide Integrated Circuit Symposium, San Diego, 1981.
18. Bosch, B. G., Gigabit electronics — a review, *Proc. IEEE,* 67, 340, 1979.
19. Derksen, R. H., Rein, H. M., and Vathke, J., Integrated bipolar master slave D-flip-flop with multiplexing capability for Gbit/s operation, *Electron. Lett.,* 20, 628, 1984.
20. Clawin, D. and Langmann, U., Monolithic multi gigabit/s silicon decision circuit for applications in fibre-optic communication systems, *Electron. Lett.,* 20, 471, 1984.
21. Hughes, J. B., Coughlin, B. C., Harbott, R. G., van der Hurk, T. H. J., van den Bergh, B. J., A versatile ECL multiplexer IC for the Gbit/s range, *IEEE J. Solid State Circuits,* 14, 812, 1979.
22. Sakai, T., Konaka, S., Kobayashi, Y., Suzuki, M., and Kawai, Y., Gigabit logic bipolar technology: advanced super self-aligned process technology, *Electron. Lett.,* 19, 283, 1983.
23. Rein, H. M., Daniel, D., Derksen, R., Langmann, U., and Bosch, B. G., Design and implementation of a Gbit/s bipolar multiplexer IC, in Symp. ESSCIRC, Lausanne, 1983.
24. Ankri, D., Scavennee, A., and Vivier, C., GaAlAs-GaAs bipolar transistors for high speed digital circuits, Paper 18, in Res. Abstr. GaAs IC Symp., San Diego, 1981.
25. Su, L. M., Grote, N., and Schmitt, F., Diffused planar InP bipolar transistor with a cadmium oxide film emitter, *Electron. Lett.,* 20, 716, 1984.
26. Nogami, M., Hirachi, Y., and Ohta, K., Present state of microwave GaAs devices, *Microelectron. J.,* 13, 29, 1982.
27. Gallager, J. J., InP: a promising material for EHF semiconductors, *Microwaves,* 21, 77, 1982.
28. Advances in GaAs IC's highlighted, *Microwave Systems News,* 12, 45, 1982.
29. Kasahara, K., Sugimoto, M., Nomura, H., and Suzuki, S., Integrated PINFET optical receiver with high-frequency InP-MISFET, *Electron. Lett.,* 19, 905, 1983.
30. Miura, S., Wada, O., Hamaguchi, H., Ito, M., Makiuchi, M., Nakai, K., and Sakurai, T., A monolithically integrated AlGaAs/GaAs p-i-n/FET photoreceiver by MOCVD, *IEEE Electron Device Lett.,* 4, 375, 1983.
31. Chen, D. R., Decker, D. R., Petersen, W. C., and Gupta, A. K., MMIC linear amplifier design and fabrication techniques, *Microwave J.,* 24, 39, 1981.
32. Liechti, C. A., Balduin, G. L., Gowen, E., Joly, R., Namjoo, M., and Podell, A. F., A GaAs MSI word generator operating at 5 Gbit/s data rate, *IEEE Trans. Electron Devices,* 29, 1094, 1982.
33. Gloanec, M., Jarry, J., and Nuzillat, G., GaAs digital integrated circuits for very high-speed frequency division, *Electron. Lett.,* 17, 763, 1981.
34. Hanjo, K. and Takayama, Y., GaAs FET ultrabroad band amplifiers for Gbit/s data rate systems, *IEEE Trans. Microwave Theory Tech.,* 29, 629, 1981.
35. Yamamoto, R. and Higashisaka, A., High speed GaAs digital integrated circuit with clock frequency of 4.1 GHz, *Electron. Lett.,* 17, 291, 1981.
36. Cathelin, M., Durand, G., Garant, M., and Rocchi, M., 5 GHz binary frequency division on GaAs, *Electron. Lett.,* 16, 535, 1980.

37. Strid, E. W., Gleason, K. R., and Addis, J., A DC-12 GHz monolithic GaAs FET distributed amplifier, Paper 47, in Res. Abstr. GaAs IC Symp., San Diego, 1981.

38. Liechti, C. A., Joly, R., and Namjoo, M., GaAs integrated circuits for error-rate measurement in high speed digital transmission systems, *IEEE J. Solid State Circuits,* 18, 402, 1983.

39. Ohta, N. and Takada, T., High-speed GaAs SCFL monolithic integrated decision circuit for Gbit/s optical repeaters, *Electron. Lett.,* 19, 983, 1983.

40. Research Abstract of the IEEE GaAs IC Symposium, Phoenix, 1983.

41. Hasegawa, H., Sawada, T., and Ishii, K., N-channel enhancement InP MISFET's using high-quality anodic $Al_2O_3$ films, Paper 19, in Res. Abstr. GaAs IC Symp., San Diego, 1981.

42. Lee, B., Lee, S. J., Hou, D., Miller, D. L., Andersen, R. J., and Sheng, N. H., High-speed frequency dividers using GaAs/GaAlAs high-electron-mobility transistors, *Electron. Lett.,* 20, 217, 1984.

43. Gheewaler, T. R., Josephson logic devices and circuits, *IEEE Trans. Electron Devices,* 27, 1857, 1980.

44. Barabas, U., 16 Gbit/s multiplexer experiment, *Electron. Lett.,* 14, 524, 1978.

45. Bahl, I. J. and Trived, D. K., A designer's guide to microstrip line, *Microwaves,* 17, 174, 1977.

46. Anders, P. and Arndt, F., Microstrip discontinuity capacitances and inductances for double steps, mitred bends with arbitrary angle, and asymmetric right-angle bends, *IEEE MTT,* 28, 1213, 1980.

47. Ch'en, D. R. and Decker, D. R., MMIC's: the next generation of microwave components, *Microwave J.,* 23, 67, 1980.

48. Ch'en, D. R., Decker, D. R., Petersen, W. C., and Gupta, A. K., MMIC linear amplifier design and fabrication techniques, *Microwave J.,* 24, 39, 1981.

49. Sergent, J. E., Understand the basics of thick film technology, *EDN,* 26, 341, 1981.

50. Sergent, J. E., Design-decision nuances forge thick-film hybrids, *EDN,* 26, 129, 1981.

51. Sergent, J. E., Thin-film hybrids provide an alternative, *EDN,* 26, 141, 1981.

52. Harper, C. A., *Handbook of Thick Film Hybrid Microelectronics,* McGraw-Hill, New York, 1974.

Chapter 8

OPTOELECTRONIC RECEIVERS

Wolfgang Albrecht and Clemens Baack

TABLE OF CONTENTS

I.    Introduction.................................................................................120

II.   Different Amplifier Concepts.........................................................120

III.  Various Transistor Stages.............................................................120
      A.   Shunt Feedback Stage .........................................................120
      B.   Series Feedback Stage..........................................................122
      C.   Bipolar Transistor in a Common Base Configuration ...................123
      D.   MESFET Stage....................................................................124

IV.   Structure of BB OERs.................................................................126
      A.   OERs of TIT.......................................................................127
           1.   TIT Amplifiers with Shunt Feedback Input Stage ...............127
           2.   TIT Amplifiers with Common Base Input Stage .................128
      B.   OERs of HIT ......................................................................129
           1.   HIT Amplifier with Series Feedback Input Stage .............130
           2.   HIT Amplifier with MESFET Input Stage ........................130
           3.   HIT Amplifier with Cascode Circuit ...............................131

V.    Noise Behavior of OERs ...............................................................133
      A.   Noise Behavior of TIT Amplifiers ..........................................136
           1.   Noise Behavior of TIT Amplifiers with Shunt Feedback
                Input Stage ...................................................................136
           2.   Noise Behavior of TIT Amplifiers with Common Base
                Input Stage ...................................................................136
      B.   Noise Behavior of HIT Amplifiers .........................................138
           1.   Noise Behavior of HIT Amplifiers with Series Feedback
                Input Stage ...................................................................138
           2.   Noise Behavior of HIT Amplifiers with MESFET
                Input Stage ...................................................................138
           3.   Noise Behavior of HIT Amplifiers with Cascode Circuit ......139
      C.   Receiver Sensitivity of Various OERs......................................139
      D.   Remarks on Noise Minimization.............................................142

VI.   Main Amplifier...........................................................................143
      A.   Three-Stage Main Amplifier Consisting of Series Feedback and
           Shunt Feedback Stages ........................................................143
      B.   Three-Stage Main Amplifier Consisting of Series Feedback
           Stages.............................................................................144

References ........................................................................................145

# I. INTRODUCTION

Optoelectronic receivers (OERs) consist of a photodiode followed by an amplifier. The photodiode converts the light signals transmitted by the optical fibers into electrical signals, which are then amplified (Chapter 2, Figure 1). Chapter 6 deals with the photodiodes in more detail. This chapter covers the design of broadband (BB) amplifiers. It also deals with the design of BB OERs using various photodiodes available today and in the near future.

# II. DIFFERENT AMPLIFIER CONCEPTS

Two principle structures can be used for preamplifiers of OERs: the transimpedance type (TIT) and the high impedance type (HIT).[1] The basic circuit diagrams for both concepts are shown in Figure 1.

In the case of the TIT the photodiode operates on a preamplifier with a low input impedance, which is achieved, e.g., by means of feedback (resistor $R_F$).

In the case of the HIT the photodiode operates on a preamplifier with a very high input impedance. Because of the time constant formed by $R_B$ and the capacities of the photodiode and the preamplifier, the circuit acts as an integrator for the signal current in the interesting frequency range. To compensate this frequency-dependence, an equalizer has to be incorporated.

To implement OERs in the gigabit per second (Gb/sec) range, BB amplifiers are necessary. The following sections will deal in greater detail with the four basic components used in the construction of BB amplifiers of both types:

- Shunt feedback stage with a bipolar transistor
- Series feedback stage with a bipolar transistor
- Bipolar transistor in common base configuration
- Metal semiconductor field effect transistor (MESFET) in common source configuration

The individual transistor stages are investigated in detail, because the thumb rules valid for low frequencies can not be used in the frequency range considered here.

# III. VARIOUS TRANSISTOR STAGES

## A. Shunt Feedback Stage

Figure 2 shows the basic circuit diagram for a shunt feedback transistor stage, and the equivalent circuit diagram of this stage for low frequencies. $R_b$, $R_1$, and $I_e$ describe the transistor at low frequencies. With the emitter series resistance $R_e$, transconductance $g_m$, and current gain $h_{FE}$ of the transistor, the following equations are valid:

$$R_1 = R_e \cdot h_{FE}; \quad I_e = R_1 \, g_m \, I_{in}$$

$$\text{with } R_F \gg (R_b + R_1) \text{ and } R_L > R_F$$

The following applies for the transient response, input, and output impedances:

$$\left| \frac{U_{out}}{I_{in}} \right| \approx R_F; \quad Z_{in} \approx \frac{R_b}{h_{FE}} + R_e; \quad Z_{out} \approx \frac{R_F}{h_{FE}}$$

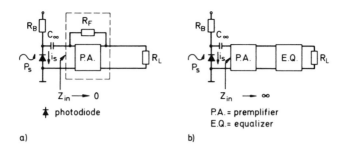

FIGURE 1.    Basic circuit diagram for the two amplifier concepts.
(a) TIT: (b) HIT.

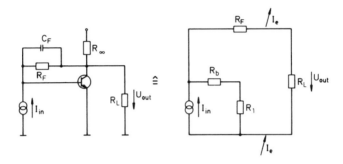

FIGURE 2.    Basic circuit diagram for a shunt feedback stage and its
equivalent circuit diagram for low frequencies.

The following calculations and investigations have been carried out using the NE 644 microwave transistor (NEC) with the following specifications.

- Collector current           $I_c$   = 8 mA
- Transit frequency         $f_T$   = 12 GHz
- Current gain              $h_{FE}$ = 200
- Base series resistance     $R_b$   = 16 $\Omega$
- Emitter series resistance  $R_e$   = 3.5 $\Omega$
- Collector emitter resistance $R_{ce}$ ≈ 1M $\Omega$

With a feedback resistor of $R_F$ = 500 $\Omega$ and $R_L$ = 50 $\Omega$ in the case of low frequencies the above values become:

$$\left|\frac{U_{out}}{I_{in}}\right| \approx 500\ \Omega \quad : \quad Z_{in} \approx 4\ \Omega; \quad Z_{out} \approx 3\ \Omega$$

We shall now examine up to which frequencies the low input and output impedances are maintained. To do this the high-frequency small signal equivalent circuit of the transistor is used (see Chapter 7, Figure 13). For the calculations $R_F$ = 500 $\Omega$, load $R_L$ = 50 $\Omega$ and a variable feedback capacity $C_F$ were chosen.

Figure 3 shows the transient response $U_{out}/I_{in}$, normalized to 1 $\Omega$ as a function of the frequency.

If $C_F$ = 0, the relationship $|U_{out}/I_{in}| \approx R_F$ derived above applies to frequencies up to the gigahertz range. However, even low values for $C_F$ have a considerable influence on the behavior of the stage.

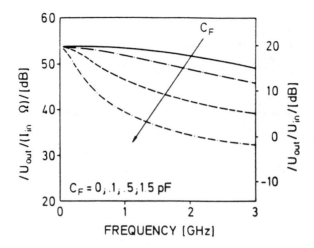

FIGURE 3.    Frequency response of the transconductance and the voltage gain.

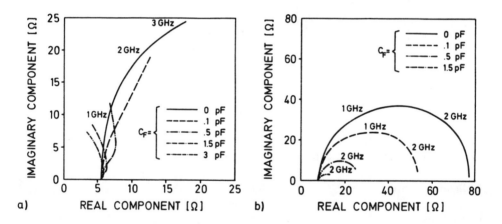

FIGURE 4.    Loci of (a) $Z_{in}$; (b) $Z_{out}$ for a shunt feedback stage, drawn for frequencies up to 3 GHz.

The loci for $Z_{in}$ and $Z_{out}$ are shown in Figure 4a and b. The values of both quantities at low frequencies agree nearly with the values of the simplified equivalent circuit diagram. However, as the frequencies increase, there are significant deviations from these values.

From these diagrams it can be seen that the capacity $C_F$ is a significant parameter for the design of a shunt feedback transistor stage.

## B. Series Feedback Stage

Figure 5 shows the basic circuit diagram for a series feedback stage and its equivalent circuit diagram for low frequencies.

For the collector emitter resistance $R_{ce} \gg 1\ \Omega$ and $R_E \gg (R_b + R_1)/h_{FE}$ we obtain the voltage gain, the input, and output impedances:

$$\left|\frac{U_{out}}{U_{in}}\right| \approx \frac{R_L}{R_E}; \quad Z_{in} \approx R_E \cdot h_{FE}; \quad Z_{out} \approx \frac{R_E}{R_e} \cdot R_{ce}$$

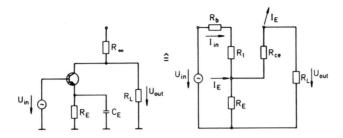

FIGURE 5. Basic circuit diagram for a series feedback stage and its equivalent circuit diagram for low frequencies.

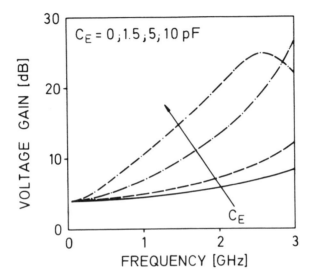

FIGURE 6. Frequency response of the voltage gain.

With a feedback resistor $R_E = 25\ \Omega$ and a load $R_L = 50\ \Omega$, we obtain at low frequencies:

$$\left|\frac{U_{out}}{U_{in}}\right| \approx 2; \quad Z_{in} \approx 5\ k\Omega; \quad Z_{out} \approx 7\ M\Omega$$

The high-frequency, small signal equivalent circuit (derived in Chapter 7) was used to calculate the frequency-dependence of these quantities.

Figure 6 shows the voltage gain as a function of the frequency and the parameter $C_E$.

Figure 7a and b shows a pronounced decrease for $Z_{in}$ and $Z_{out}$, as the frequency increases with only a slight dependence of both quantities on the parameter $C_E$.

## C. Bipolar Transistor in Common Base Configuration

Figure 8 shows a transistor in common base configuration and its equivalent circuit for low frequencies.

For the current gain, input and output impedances we obtain with $R_B = 50\ \Omega$:

$$\left|\frac{I_{out}}{I_{in}}\right| \approx 1; \quad Z_{in} \approx R_e + \frac{R_B}{h_{FE}}; \quad Z_{out} \approx R_{ce} \cdot h_{FE}$$

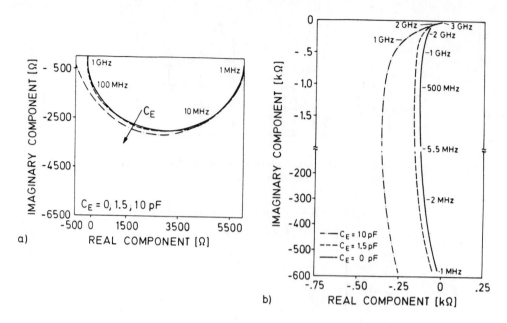

FIGURE 7.    Loci of (a) $Z_{in}$; (b) $Z_{out}$ for a series feedback stage.

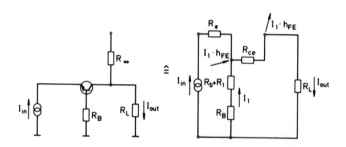

FIGURE 8.    Basic circuit diagram for a transistor in common base configuration and its equivalent circuit for low frequencies.

For low frequencies this then yields:

$$Z_{in} \approx 4 \ \Omega; \quad Z_{out} \approx 200 \ M\Omega$$

Figures 9 and 10 again show the frequency-dependence of the current gain, and the input and output impedances.

### D. MESFET Stage

Figure 11 shows the MESFET stage and the simplified equivalent circuit for low frequencies. As an example with $R_{gs} = 10 \ M\Omega$, $R_{ds} = 250 \ \Omega$, $g_m = 35 \ mS$, and $R_L = 50 \ \Omega$, we obtain the voltage gain, the input, and output impedances:

$$\left| \frac{U_{out}}{U_{in}} \right| = 1.5; \quad Z_{in} = 10 \ M\Omega; \quad Z_{out} = 250 \ \Omega$$

In the following calculations the MESFET small signal equivalent circuit of Chapter 7 is used. Figure 12 shows the voltage gain as a function of the frequency. Figure 13 depicts the loci of $Z_{in}$ and $Z_{out}$.

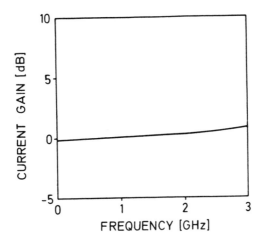

FIGURE 9.    Frequency response of the current
gain.

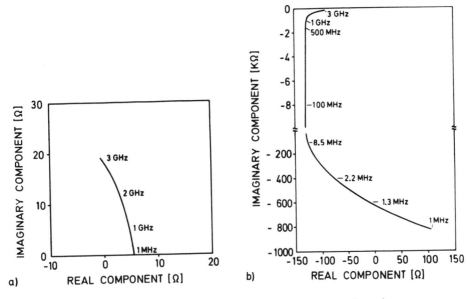

FIGURE 10.    Loci of (a) $Z_{in}$; (b) $Z_{out}$ for a common base configuration.

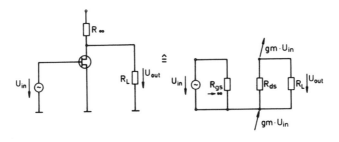

FIGURE 11.    MESFET stage and its simplified equivalent circuit
diagram for low frequencies.

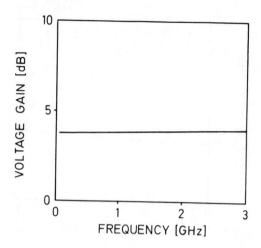

FIGURE 12.    Frequency response of the voltage gain.

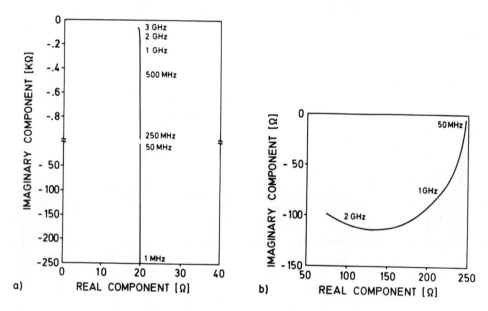

FIGURE 13.    Locus of (a) $Z_{in}$; (b) $Z_{out}$ of a MESFET stage.

In conclusion, the important properties of the considered transistor stages for low frequencies are given in Table 1.

## IV. STRUCTURE OF BB OERs

In the following, BB amplifiers will be set up using the four transistor stages described in detail in the preceding section.

Designing BB amplifiers is of importance to exploit the transistors up to their transit frequency. The maximum gain bandwidth product is only achieved if the transfer function of the amplifier is determined solely by the time constants of the transitor, and thus by the transit frequency, and not by the time constants due to parasitic elements. According to Cherry-Hooper[2] the influence exerted by parasitic elements can be reduced if the individual amplifier stages can be cascaded with the maximum possible

Table 1

PROPERTIES OF DIFFERENT TRANSISTOR STAGES FOR
LOW FREQUENCIES

| | $Z_{in}$ | $Z_{out}$ | |
|---|---|---|---|
| Shunt feedback stage | Low | Low | $\lvert U_{out}/U_{in} \rvert \approx R_F$ |
| Series feedback stage | High | Very high | $\lvert U_{out}/U_{in} \rvert \approx R_L/R_E$ |
| Common bases configuration | Very low | Very high | $\lvert I_{out}/I_{in} \rvert \approx 1$ |
| Common source configuration | Very high | High | $\lvert U_{out}/U_{in} \rvert \approx R_L g_m /$ $(1+R_L/R_{ds})$ |

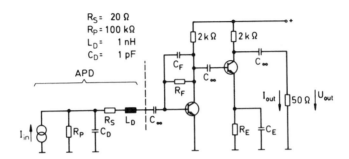

FIGURE 14.    TIT amplifier with shunt feedback input stage.

mismatching. A stage with a high output impedance should be followed by a stage with the smallest possible input impedance, and vice versa. This establishes the guidelines for combining the four stages described above, in order to set up OERs of TIT or of HIT with the maximum possible gain for a given bandwidth.

## A. OERs of TIT
In the following we will discuss two different concepts for OERs of TIT.

### 1. TIT Amplifiers with Shunt Feedback Input Stage
The TIT requires a preamplifier with the smallest possible input impedance, to ensure that the time constants formed by the capacity of the photodiode and the input impedance of the amplifier are sufficiently small and ineffectual. The shunt feedback stage shown in Figure 2 satisfies the conditions for a low input impedance, offers a low output impedance, and thus the following stage should have a high input impedance. This condition is satisfied by a series feedback stage as shown in Figure 5. This results in the two-stage TIT receiver shown in Figure 14, which operates at a load of 50 Ω. (The model of the avalanche photodiode (APD) is described in Chapter 6.)

Figure 15 displays the frequency response of the current gain of the two-stage amplifier for various $R_F C_F$ and $R_E C_E$ combinations.

Figure 16a shows the loci for $Z_{out}$ of the first stage together with $Z_{in}$ of the second stage of the amplifier (corresponding to curve 3, Figure 15). The criterion of mismatching is satisfied very effectively in the case of low frequencies, i.e., a high input impedance follows the low output impedance of the first stage. In the case of high frequencies the situation changes, as indicated in Figure 16b. The input impedance of the second stage is rapidly reduced as the frequency increases, and in the gigahertz range it assumes the magnitude of the output impedance of stage 1. At high frequencies $Z_{in}$ and $Z_{out}$ are approximately 50 Ω.

Figure 17 depicts the input and output impedance of the complete TIT amplifier, corresponding to curve 3 (Figure 15) for a 3-dB bandwidth of 2 GHz. As shown in

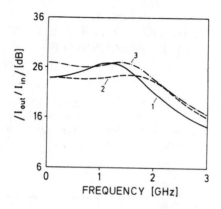

| curve | $R_F$ [Ω] | $C_F$ [pF] | $R_E$ [Ω] | $C_E$ [pF] | $I_{out}/I_{in}$ [dB] | 3 dB bandwidth [GHz] |
|---|---|---|---|---|---|---|
| 1 | 500 | 0 | 25 | 0 | 24 | 2.0 |
| 2 | 500 | 0.22 | 25 | 5.3 | 24 | 2.3 |
| 3 | 840 | 0.15 | 31 | 4.2 | 27 | 2.0 |

FIGURE 15.    Frequency response of the current gain for various $R_F C_F$ and $R_E C_E$ combinations.

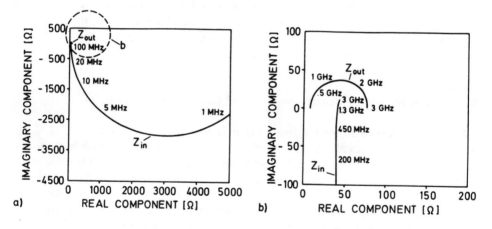

FIGURE 16.    Loci of $Z_{out}$ of the first stage and $Z_{in}$ of the second stage; Figure b shows a detail from Figure a.

Figure 17a, the photodiode is loaded by an impedance with a real component of < 50 Ω for all frequencies.

## 2. TIT Amplifiers with Common Base Input Stage

A bipolar transistor in a common base configuration (Figure 18) can also be used as input stage for a TIT amplifier. The common base configuration has only a low input impedance and, in contrast to the shunt feedback stage, a high output impedance. To achieve the mismatching condition, a shunt feedback stage follows the input stage.

Figure 19 displays the frequency response of the current gain for $R_F = 400$ Ω, $R_E = 11$ Ω, $C_F = 0.3$ pF, and $C_E = 4.6$ pF. These values are optimized for maximum bandwidth. Given a current gain of 26.5 dB, we achieve a 3-dB bandwidth of 2.4 GHz. A

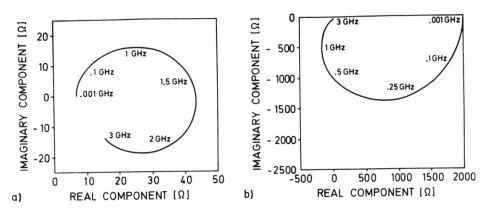

FIGURE 17. Loci for input (a) and output (b) impedances of the TIT amplifier, corresponding to curve 3, Figure 15.

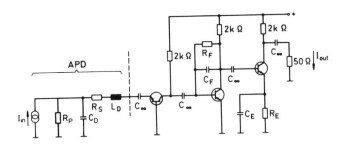

FIGURE 18. TIT amplifier with transistor in common base config-uration.

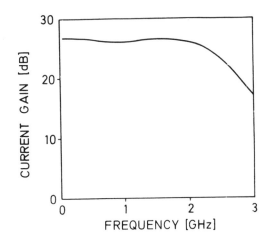

FIGURE 19. Frequency response of the current gain.

gain bandwidth product of 51 GHz can be attained, and with the TIT amplifier as shown in Figure 14, a maximum gain bandwidth product of 45 GHz (curve 3, Figure 15) was possible.

## B. OERs of HIT

Three different concepts are discussed below for OERs of HIT.

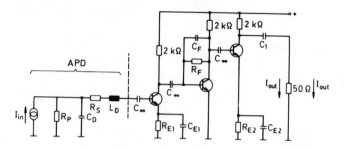

FIGURE 20.    HIT amplifier with series feedback input stage.

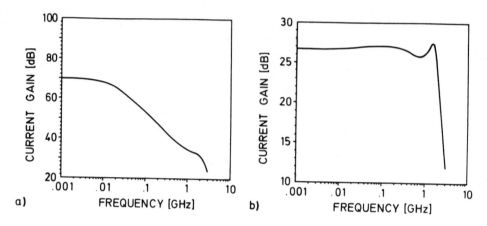

FIGURE 21.    Frequency response of the current gain. (a) Before equalization; (b) after equalization.

## 1. HIT Amplifier with Series Feedback Input Stage

The HIT requires a preamplifier with a high input impedance. The series feedback stage as shown in Figure 5 satisfies this condition. The series feedback also offers high output impedance, and thus the subsequent stage should have a low input impedance. This requirement is satisfied, e.g., by a shunt feedback stage as shown in Figure 2. Once again, as described in the previous section, a series feedback stage serves as the third stage (Figure 20).

The frequency response of the current gain shown in Figure 21a is obtained for the values $R_{E1} = 26\ \Omega$, $R_F = 500\ \Omega$, $RE_2 = 26\ \Omega$, $C_1 \to \infty$, $C_{E1} = 5$ pF, $C_F = 0.25$ pF, and $C_{E2} = 3.3$ pF. For use at high bit rates this amplifier must be equalized. This is achieved by varying the values for $R_{E1}$, $R_F$, $R_{E2}$, $C_1$, $C_{E1}$, $C_F$, and $C_{E2}$. With $R_{E1} = 26\ \Omega$, $R_F = 200\ \Omega$, $R_{E2} = 240\ \Omega$, $C_{E1} = 5$ pF, $C_F = 1$ pF, $C_{E2} = 7.5$ pF, and $C_1 = 0.8$ pF, we obtain the frequency response shown in Figure 21b. With a gain of 27 dB, a 3-dB bandwidth of 2 GHz is achieved. For completeness the input impedance of the HIT amplifier is shown in Figure 22.

## 2. HIT Amplifier with MESFET Input Stage

A MESFET can also be used as an input stage (Figure 23) in place of the bipolar transistor of Figure 20. With $R_F = 500\ \Omega$, $R_E = 25\ \Omega$, $C_1 \to \infty$, $C_F = 0.25$ pF, and $C_E = 3.3$ pF, Figure 24a shows the frequency response of the current gain. Equalization of this integrator is achieved by varying the feedback time constants and $C_1$. With $R_F = 790\ \Omega$, $R_E = 900\ \Omega$, $C_F = 0.1$ pF, $C_E = 1.5$ pF, and $C_1 = 2.5$ pF, a current gain of 26 dB as shown in Figure 24b and a 3-dB bandwidth of 2.3 GHz are obtained.

Figure 25 shows the locus of the input impedance of the HIT amplifier with MES-

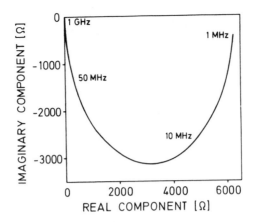

FIGURE 22.   Input impedance of the HIT amplifier.

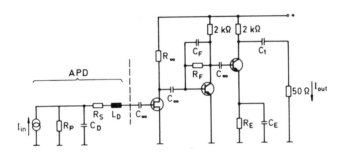

FIGURE 23.   HIT amplifier with MESFET input stage.

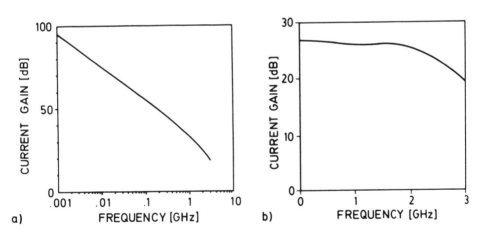

FIGURE 24.   Frequency response of the current gain. (a) Without equalization; (b) with equalization.

FET input stage, where we see in contrast to Figure 13a a large real component for low
frequencies.

### 3. HIT Amplifier with Cascode Circuit

The second stage in Figure 23 now is replaced by a bipolar transistor in a common
base configuration (see also Figure 8). This stage does display a high output impedance, and for this reason a shunt feedback stage is used as the third stage (Figure 26).
Figure 27 displays the equalized frequency response of the current gain.

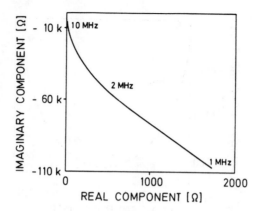

FIGURE 25.    Locus of the input impedance of a
HIT amplifier with MESFET input stage.

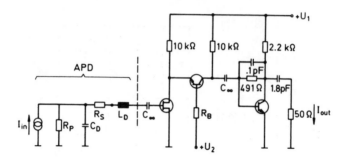

FIGURE 26.    HIT amplifier in cascode configuration.

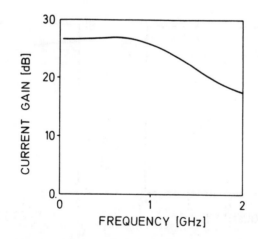

FIGURE 27.    Frequency response of the current
gain.

Limitations are, however, imposed on the BB nature of this circuit. As shown in
Figure 27, given a current gain of 26.5 dB it is not possible to improve on a 3-dB
bandwidth of 1.3 GHz. Due to the good noise properties that it offers (see next section)
we shall devote more attention to this circuit concept.

Figure 28 shows the locus for the input impedance. Comparison with Figures 13a

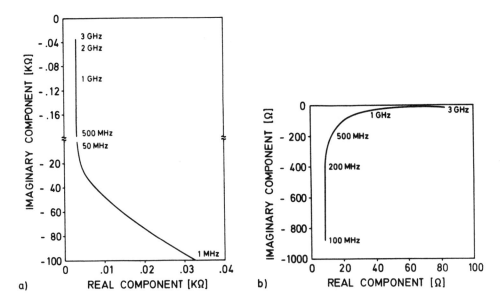

FIGURE 28. Loci of (a) $Z_{in}$; (b) $Z_{out}$ of a HIT amplifier in cascode configuration.

Table 2
CURRENT GAIN AND 3-dB
BANDWIDTH OF VARIOUS
AMPLIFIERS

| | Input stage | GI (dB) | B (GHz) |
|---|---|---|---|
| TIT | Shunt feedback | 27 | 2.0 |
| TIT | Base configuration | 26.5 | 2.4 |
| HIT | Series feedback | 27 | 2.0 |
| HIT | MESFET | 26 | 2.3 |
| HIT | MESFET (cascode) | 26.5 | 1.3 |

and 25 shows in particular the dependence of the real component of the input impedance of the MESFET stage with a load of 50 Ω (Figure 13a), with a shunt feedback stage (Figure 25), and in this case with a common base configuration (Figure 28).

Finally, the current gain and bandwidth of the five amplifier concepts are given in Table 2.

## V. NOISE BEHAVIOR OF OERS

In addition to the frequency behavior of the amplifier, as examined in the last section, the noise behavior of individual circuits is of major importance for the design of transmission systems. The noise behavior determines the sensitivity of the OER, which in turn has an influence on the repeater spacing. The signal-to-noise ratio (SNR) at the output of the OER is given as follows:

$$\frac{S}{N} = \frac{P_S}{P_{NS} + P_{ND} + P_{NR} + P_{NA}}$$

$P_S$    Signal power

$P_{NS}$    Signal-dependent noise of the photodiode

$P_{ND}$    Dark current noise of the photodiode

$P_{NR}$    Thermal noise of the diode biasing resistor $R_p$

$P_{NA}$    Amplifier noise

(1)

The signal power is calculated as follows:

$$P_S = R_L \int_0^\infty |i_s(f)|^2 \cdot |GI_{res}(f)|^2 \, df$$

$$= \left(\frac{\eta e \cdot M}{h \cdot \nu}\right)^2 R_L \int_0^\infty |p(f)|^2 \cdot |GI_{res}(f)|^2 \, df \qquad (2)$$

p(f)       Absolute power density of the received light signal
$GI_{res}(f)$    Spectral current gain of the amplifier
$R_L$        Load resistor

The noise properties of the photodiode are described in detail in Chapter 6 and in Reference 1.

The resulting signal-dependent noise power is

$$P_{NS} = 2e \frac{\eta e \overline{P}}{h \cdot \nu} F_P(M) \, M^2 R_L \int_0^\infty |GI_{res}(f)|^2 \, df$$

$\overline{P}$    Received average optical power

(3)

The dark current noise power is

$$P_{ND} = 2eI_{do}F_d(M) \, M^2 R_L \int_0^\infty |GI_{res}(f)|^2 \, df \qquad (4)$$

The thermal noise power of the biasing resistor is

$$P_{NR} = \frac{4kT}{R_p} R_L \int_0^\infty |GI_{res}(f)|^2 \, df \qquad (5)$$

The amplifier noise power $P_{NA}$ is examined in more detail below.

From Equation 5 for $P_{NR}$ it is apparent that the biasing resistor $R_P$ should be as large as possible. However, its magnitude is limited by the required dynamic range and the available diode voltage. Under practical conditions a value of $R_P$ can usually be selected that permits the noise term $P_{NR}$ to be ignored.

Figure 29 shows the signal power and the various noise powers dependent on the multiplication factor M of the photodiode. In a logarithmic scale, the SNR may be read directly as the difference between the curves for $P_S$ and $P_{NRES}$. The curves apply for a Ge-APD with a capacity of $C_D = 1$ pF and a dark current $I_D = 0.4$ $\mu$A followed by a TIT amplifier, as shown in Figure 14 with a 3-dB bandwidth of 2 GHz. The received light power amounts to 1 $\mu$W.

The signal power $P_S$ at the output to the optoelectronic amplifier increases proportional to $M^2$. The signal-dependent noise power $P_{NS}$, and in especially the dark current noise power $P_{ND}$ of the photodiode increase at a higher rate than $M^2$. The amplifier noise power $P_{NA}$ and the noise power $P_{NR}$ of the diode biasing resistor are independent of M. The optimum multiplication factor $M_{OPT}$ is the value of M resulting in the maximum SNR.

The equivalent noise circuits for the transistors, as shown in Chapter 7, are used in order to investigate the noise behavior of the individual amplifer concepts. The advan-

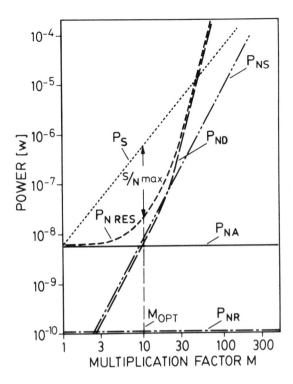

FIGURE 29.   Signal and noise powers at the output of the
amplifier; received light power: 1 μW.

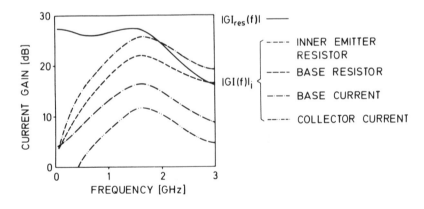

FIGURE 30.   Current gain of the amplifier and transmission functions for individual
noise sources of the first transistor stage.

tage of these equivalent circuits is that the noise sources are assumed at that point
where they are also to be found in the case of real components. This fact must be taken
into account in order to achieve precise calculations, because each noise source must
be weighted according to the transmission function between the location of the noise
source and the load $R_L$.

Figure 30 shows the frequency response of the signal current gain $GI_{res}$ (f) of the TIT
amplifier used as shown in Figure 14, together with the frequency responses for the
current gain $GI$ (f)$_i$, by means of which the various noise sources are weighted.

The noise power $P_{NA}$ at the amplifier output is calculated:

$$P_{NA} = \sum_{i=1}^{n} [|i_{reff\ i}|^2\ R_L \int_0^{\infty} |GI(f)_i|^2\ df]$$

$$+ \sum_{K=1}^{m} \left[\frac{|u_{reff\ K}|^2}{R_L} \int_0^{\infty} |GV(f)_K|^2\ df\right] \qquad (6)$$

In this calculation $i_{reff}$ describes the noise current sources and $u_{reff}$ the noise voltage sources of the transistors, which are weighted with the current and voltage gain GI and GV. Admittedly, this formula requires a vast amount of calculation, but it is the only feasible way of obtaining an exact description of the noise properties. In the derivation of this formula it was assumed that the noise sources supply white noise and are not correlated with one another.

## A. Noise Behavior of TIT Amplifiers
In the following section the noise behavior of the two TIT receivers described in Section III.A is investigated.

### 1. Noise Behavior of TIT Amplifiers with Shunt Feedback Input Stage
Figure 14 shows the TIT amplifier with shunt feedback input stage.

First the influence of the capacity $C_D$ and inductance $L_D$ of the photodiode on the noise behavior is examined.

Figure 31a shows the complete spectral noise power density of the amplifier. The parameter is the diode capacity $C_D$. The associated curves for the current gain of the amplifier are (compare curve 3, Figure 15) obtained from Figure 31b. Here it is clear that the spectral noise power densities possess different frequency-dependencies to that of the current gain $GI_{res}$. For this reason the amplifier noise cannot be simulated by an equivalent noise resistance at the amplifier input.

The bandwidth decreases significantly as the capicitor $C_D$ increases, while the noise also increases significantly. Therefore low capacity photodiodes are an essential feature in the design of low-noise and BB amplifiers.

Corresponding investigations were carried out for the connection inductance $L_D$ of the photodiode. As can be seen in Figure 32a, the noise behavior of the amplifier is hardly influenced at all by the inductance. The frequency response of the current gain is included in Figure 32b. This can even be improved slightly by an inductance.

The following table shows which noise sources make a decisive contribution to the total amplifier noise (see Figure 13, Chapter 7):

| | | | |
|---|---|---|---|
| 1. | Collector current | 1st stage | 2.33 nW |
| 2. | Internal base series resistor | 1st stage | 1.61 nW |
| 3. | Base current | 1st stage | 1.42 nW |
| 4. | Feedback resistor | 1st stage | 0.815 nW |
| 5. | External base series resistor | 1st stage | 0.129 nW |
| 6. | Emitter series resistor | 1st stage | 0.126 nW |

The noise power applies for a bandwidth of 2 GHz. As expected, with this type of amplifier the major noise sources are found in the first stage. Apart from the noise of the feedback resistor, all other noise sources are determined by the properties of the transistors.

### 2. Noise Behavior of TIT Amplifiers with Common Base Input Stage
The optoelectronic TIT receiver with a bipolar transistor in a common base configuration is shown in Figure 18.

Figure 33 describes the noise behavior. The spectral noise power density is higher

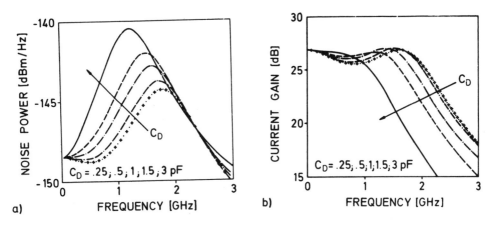

FIGURE 31.   Frequency responses $L_D = 1$ nH. (a) Spectral noise power density; (b) current gain GI.

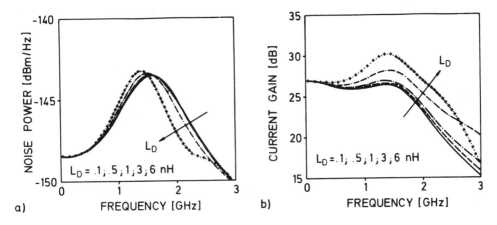

FIGURE 32.   Frequency responses $C_D = 1$ pF. (a) Spectral noise power density; (b) current gain GI.

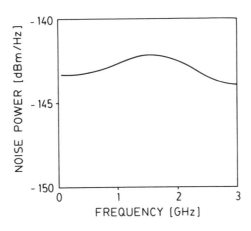

FIGURE 33.   Spectral noise power density of a TIT amplifier in common base configuration.

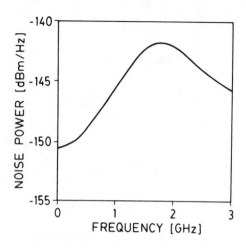

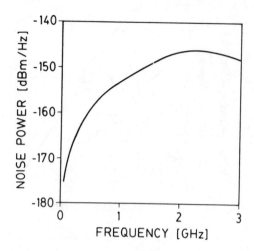

FIGURE 34.    Spectral noise power density of an equalized HIT amplifier with series feedback input stage.

FIGURE 35.    Spectral noise power density of an equalized HIT amplifier with MESFET input stage.

than in the case of the TIT amplifier with shunt feedback input stage. The common base configuration operates at a low impedance load (input impedance of the shunt feedback stage), and no amplification of the signal power can therefore take place; only additional noise power occurs. Although a high gain bandwidth product can be achieved using a TIT amplifier with base input stage, such amplifiers should not be used in OERs due to their unfavorable noise properties.

## B. Noise Behavior of HIT Amplifiers

In the following section the noise behavior of the three HIT receivers described in Section III.B will be examined.

### 1. Noise Behavior of HIT Amplifiers with Series Feedback Input Stage

The optoelectronic HIT receiver with series feedback input stage is displayed in Figure 20.

The spectral noise power density is shown in Figure 34.

The following table shows which noise sources make a decisive contribution towards the amplifier noise:

| | | | |
|---|---|---|---|
| 1. | Collector current | 1st stage | 1.93 nW |
| 2. | Base current | 1st stage | 1.72 nW |
| 3. | Internal base series resistor | 1st stage | 1.50 nW |
| 4. | Feedback resistor | 1st stage | 0.58 nW |
| 5. | Internal base series resistor | 2nd stage | 0.27 nW |
| 6. | Feedback resistor | 2nd stage | 0.20 nW |

In this circuit the noise sources of the first stage are no longer the sole determining factor, significant noise components also being produced in the second stage. In contrast to the TIT amplifier, in the case of HIT amplifiers we must take the noise sources of all stages into account.

### 2. Noise Behavior of HIT Amplifiers with MESFET Input Stage

The optoelectronic HIT receiver with MESFET input stage is shown in Figure 23.

The calculations were carried out for a MESFET with a leakage current of only 10 nA[4]. Figure 35 shows very low values for the spectral noise power density at low fre-

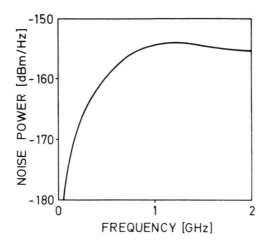

FIGURE 36.    Spectral noise power density of an
equalized HIT amplifier with cascode circuit.

quencies. Resulting from equalization, the noise components of the following stage become increasingly important, as the frequency becomes higher. This can also be seen from the following list of the major noise sources:

1. Source resistor MESFET          1st stage   0.49 nW
2. Drain resistor                        1st stage   0.27 nW
3. Collector current                   2nd stage 0.22 nW
4. Internal base series resistor    2nd stage 0.17 nW
5. Base current                          2nd stage 0.16 nW
6. Feedback resistor                   2nd stage 0.10 nW

The list shows that a main part of the noise power is contributed by the second stage. At this point it should again be pointed out that one should not only take into account the noise source of the first stage when considering the noise aspect for BB HIT amplifiers.

*3. Noise Behavior of HIT Amplifiers with Cascode Circuit*
Figure 26 shows an optoelectronic, HIT receiver with cascode circuit configuration. Although no particularly large bandwidth can be achieved with this type of amplifier we shall still examine its noise behavior. Figure 36 shows the noise power as a function of the frequency. In the case of low frequencies this circuit produces very little noise.

Even at increasing frequencies it remains below the noise power level of all the types of amplifiers examined so far. The main noise sources are as follows:

1. Source impedance MESFET   1st stage   0.125 nW
2. Base current                          2nd stage 0.040 nW
3. Base current                          3rd stage 0.037 nW
4. Feedback resistor                   3rd stage 0.035 nW
5. Collector current                   2nd stage 0.026 nW
6. Collector current                   3rd stage 0.024 nW

The integration bandwidth for determining the noise power in this case was 1.5 GHz. Compared with the HIT amplifier with MESFET described above, the internal base series resistor of the second transistor produces less noise in the common base configuration than in the common emitter configuration.

## Table 3
## SPECIFICATIONS OF THE PHOTODIODES EXAMINED

|  | Ge-APD | III-V-PIN-Diode | III-V-APD |
|---|---|---|---|
| $\eta$ | 0.7 | 0.7 | 0.8 |
| $K_i$ | 0.8 | — | 0.3 |
| $I_{DO}(A)$ | $0.4 * 10^{-6}$ | $10^{-9}$ | $0.5 * 10^{-9}$ |
| $C_D(pF)$ | 1 | 0.25 | 0.5 |

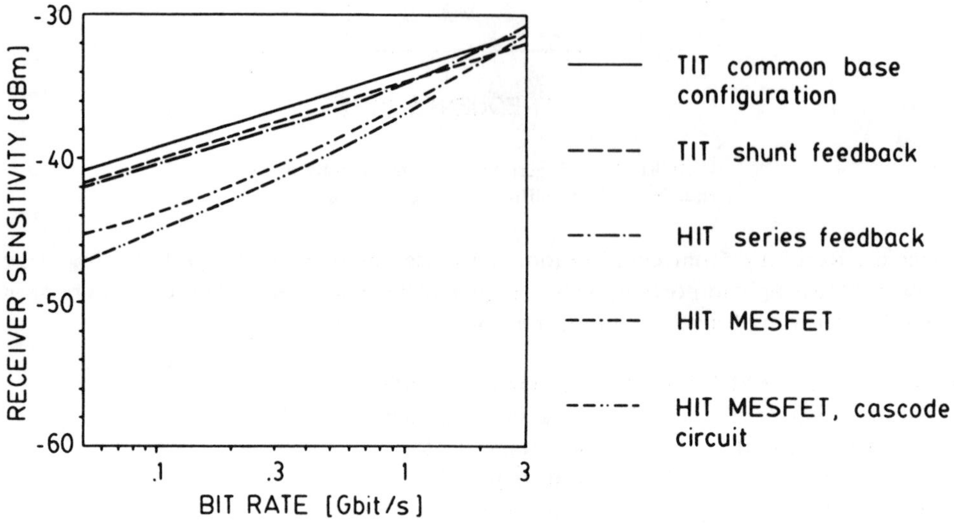

FIGURE 37.     Receiver sensitivity of OERs consisting of a Ge-APD and various preamplifiers.

## C. Receiver Sensitivity of Various OERs

Below we will examine how the individual amplifier concepts, taken together with various photodiodes, influence the receiver sensitivity of the OER. Our investigations will concentrate on OERs with Ge-APDs, with III-V-PIN photodiodes, and with future III-V APDs.[3] Table 3 contains the important specifications for these components.

The transmission and noise behaviors of the five BB amplifiers were discussed in Section III and in Section IV.A and B, respectively.

Random NRZ rectangular light pulses at the input of the OER are assumed. To investigate the receiver sensitivity for a bit error rate of $10^{-9}$ the following method is applied: each amplifier is optimized for the required bandwidth of 1 GHz (HIT with cascode circuit) or 2 GHz (all other amplifiers). For calculating the amplifier noise power as a function of the bit rate, the noise bandwidth is chosen to the 1.6-fold Nyquist frequency, without changing the circuit of the amplifiers. This method is permissible as a means of comparing various amplifier concepts and it is commonly used in literature, although in practice the circuit should be optimized for the respective bit rate.

First the Ge-APD (Table 3) is combined with all amplifiers. Figure 37 shows the receiver sensitivity as a function of the bit rates.

If Ge-APDs are the only available photodiodes, HIT amplifiers with cascode circuit should be used for low bit rates (< 1 Gb/sec), and HIT amplifiers with MESFET input stage or TIT amplifiers with shunt feedback input stage should be used for higher bit rates (> 1 Gb/sec).

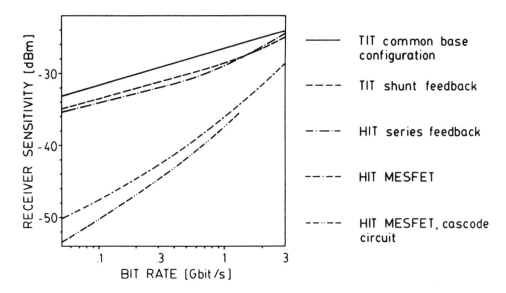

FIGURE 38.    Receiver sensitivity of OERs consisting of a III-V-PIN photodiode and various amplifiers.

We shall now turn to the III-V-PIN photodiode currently available, which is combined with all five types of amplifiers. Figure 38 shows the receiver sensitivity as a function of the bit rate.

If III-V-PIN photodiodes are to be used, a HIT amplifier with cascode circuit should be used for low bit rates (< 1 Gb/sec), and a HIT amplifier with MESFET input stage for high bit rates (> 1.0 Gb/sec).

Finally, a III-V-APD, which already exists as a laboratory device, is combined with all five kinds of amplifiers. Figure 39 shows the receiver sensitivities as a function of the bit rate.

Improvement in the sensitivity when using a III-V-APD is considerable when compared with a Ge-APD and particularly in comparison with a III-V-PIN photodiode. Concerning the selection of amplifiers for the various bit rates, the same details are valid as mentioned for the III-V-PIN photodiode. Figure 40 summarizes the concepts offering the lowest noise, from Figures 37, 38, and 39.

For all bit rates the most sensitive OERs can be set up using III-V-APDs (curves 1 to 3). For low bit rates (< 1 Gb/sec) a HIT amplifier with cascode arrangement should be used. For higher bit rates HIT amplifiers with MESFET input stage or TIT amplifiers with shunt feedback input stage should be implemented.

Up to now however, only Ge-APDs and III-V-PIN photodiodes have been available. In accordance with Figure 40, PIN photodiodes and HIT amplifiers with cascode circuit arrangement (curve 4) should be used for low bit rates (< 1.0 Gb/sec), and Ge-APDs and HIT amplifiers with MESFET input stage (curve 5) or TIT amplifiers with shunt feedback input stage (curve 6) should be used for high bit rates.

A number of measured values are listed in Figure 40: curve 4 (PIN photodiode with HIT in cascode circuit, British telecom[5]) and curve 6 (Ge-APD with TIT and shunt feedback input stage, HHI measurements).

When deciding between HIT and TIT receivers, it should be borne in mind that the TIT receiver has a considerably higher dynamic range. Furthermore, the circuit of the HIT receiver is more complicated than that of the TIT receiver because a third transistor stage is required (compare Figures 14 and 23).

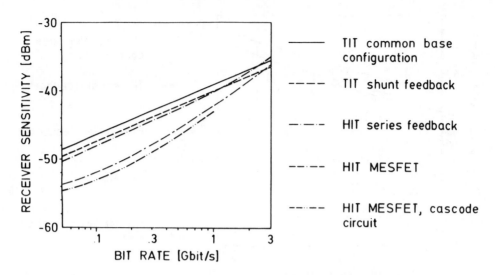

FIGURE 39.    Receiver sensitivity of OERs with a III-V-APD and different preamplifiers.

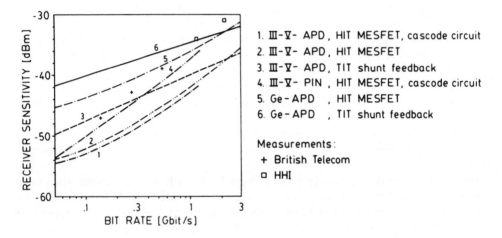

FIGURE 40.    Receiver sensitivity of OERs with a Ge-APD, a III-V-PIN photodiode, and a III-V-APD.

## D. Remarks on Noise Minimization

Using the HIT amplifier as an example, attention should be drawn to a number of items in the development of an OER.

First the influence of the MESFET leakage on the sensitivity of a PINFET receiver is examined. Figure 41 shows the sensitivity for two different leakage currents. The value of 10 μA is customary with commercially available transistors. Certain special laboratory transistors do, however, achieve values of 10 nA.[4] This difference has a noticeable effect on sensitivity, particularly in the case of low bit rates. For 50 Mb/sec, for example, a 6-dB improvement in sensitivity can be achieved. At high bit rates it is not possible to establish any influence of the MESFET leakage current, because the subsequent bipolar transistors and not only the low noise MESFET, is the important element determining the noise.

The required frequency response for a HIT amplifier can be achieved with equalization taking place prior to the first transistor stage or alternatively in the second and third transistor stages. The receiver sensitivities are different if the equalization takes place prior to the first stage or in the subsequent stages. Figure 42 displays these two

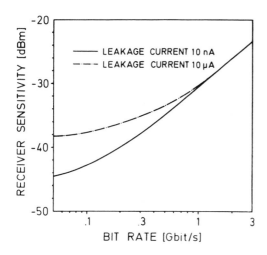

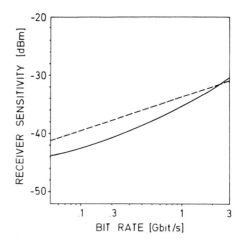

FIGURE 41.    Influence of the leakage current of the MESFET on the receiver sensitivity.

FIGURE 42.    Equalization of a HIT amplifier. Dashed line: in front of the first stage; solid line: in the subsequent stages.

cases. In the case of bit rates up to 1 GHz there is an improvement of approximately 2 dB when equalization takes place in the subsequent stages. Therefore a HIT amplifier should be equalized as near the last stage as possible in order to improve the noise behavior.

Finally we shall examine the influence of the photodiode capacity on the receiver sensitivity. In examining the frequency response we have already established that this is adversely affected by a large diode capacity. The same applies to the sensitivity, as shown in Figure 43 (see also Section IV.A.1). For all bit rates the sensitivity is adversely affected by increased capacities.

## VI. MAIN AMPLIFIER

The amplifier shown in Figure 1, Chapter 2, consists of a HIT or TIT preamplifier and a subsequent main amplifier.

In contrast to the preamplifier, the main amplifier does not influence the SNR ratio of the transmission system. In planning, therefore, attention should not be paid to the noise behavior, but to the following requirements:

1.    A 3-dB bandwidth of at least 2 GHz
2.    The input reflection factor in this range should be $< -20$ dB
3.    The output voltage dynamic range should be as large as possible

First we shall consider a main amplifier consisting of series feedback and shunt feedback stages. Subsequently an amplifier will be set up using only series feedback stages.

### A. Three-Stage Main Amplifier Consisting of Series Feedback and Shunt Feedback Stages

The basic circuit for this amplifier is shown in Figure 44. The input network of 33- and 18-$\Omega$ resistors permit matching to a 50-$\Omega$ system. This is a very simple solution, and up to the gigahertz range it displays good results. The two 50-$\Omega$ coaxial cables with lengths $l_1$ and $l_2$ are provided for frequency compensation purposes (see Chapter 7, Section II.C.1).

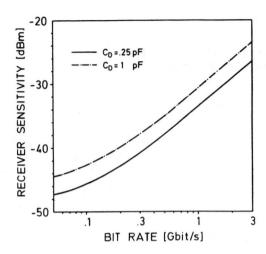

FIGURE 43.    Influence of the diode capacity $C_D$ on the receiver sensitivity.

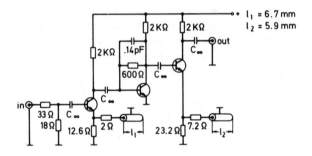

FIGURE 44.    Basic circuit for the three-stage main amplifier I.

The frequency response of this amplifier is shown in Figure 45 as a solid line. With a current gain of 27 dB, it is possible to achieve a 3-dB bandwidth of 3 GHz.

The magnitude of the reflection factor $S_{11}$ is shown in Figure 46. Up to 2 GHz it is in the order of −30 dB and below. These low reflection factors are of major significance in digital systems, where pulse distortion as a result of mismatching should be kept as low as possible.

One problem of decisive importance when implementing this circuit is found in the design of the second stage. The feedback capacity of 0.14 pF must be kept with a great deal of precision, which is very difficult. Moreover, the entire feedback network of this stage is difficult to implement in a compact design if thick film technology is used.

### B. Three-Stage Main Amplifier Consisting of Series Feedback Stages

A further possibility for setting up a main amplifier consists in connecting of three series feedback stages. The basic circuit for this amplifier is shown in Figure 47. With a current gain of 17 dB and a 3-dB bandwidth of 2 GHz (Figure 45), neither the gain nor the bandwidth of the first circuit can be achieved. The advantage of this circuit concept is that it can be very easily implemented using thick film technology.

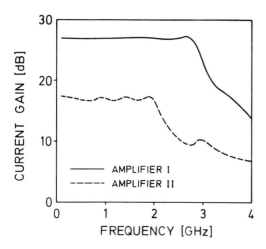

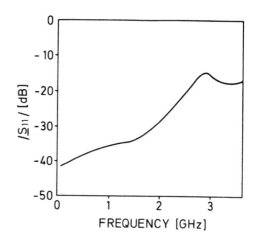

FIGURE 45.    Frequency responses of the current gain of the main amplifiers I and II.

FIGURE 46.    Magnitude of the reflection factor $\underline{S}_{11}$ of main amplifier I.

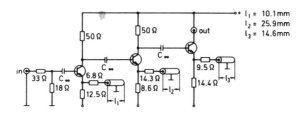

FIGURE 47.    Circuit diagram for the three-stage main amplifier II.

# REFERENCES

1. Smith, R. G. and Personick, S. D., Receiver design for optical fiber communication systems, in *Semiconductor Devices for Optical Communication*, Kressel, H., Ed., Springer-Verlag, Berlin, 1982, chap. 4.
2. Cherry, E. M. and Hooper, D. E., The design of wideband transistor feedback amplifiers, *Proc. IEEE*, 110(2), 375, 1963.
3. Muoi, T. V., Receiver design for high-speed optical fiber systems, *J. Lightwave Technol.*, 2(3), 243, 1984.
4. Smith, D. R., Hooper, R. C., and Garrett, I., Receivers for optical communications: a comparison of avalanche photodiodes with PIN-FET hybrids, *Optical Quantum Electron.*, 10, 293, 1978.
5. Smith, D. R., Hooper, R. C., Smyth, P. P., Wake, D., Experimental comparison of a Germanium avalanche photodiode and InGaAs PINFET receiver for longer wavelength optical communications systems, *Electron. Lett.*, 18(11), 453, 1982.

Chapter 9

# SIGNAL GENERATION AND REGENERATION

Bernhard Enning, Günter Heydt, Lutz Ihlenburg, and Godehard Walf

## TABLE OF CONTENTS

I. Tasks of Generation and Regeneration ................................................ 148

II. Signal Generation ........................................................................ 148
    A. High-Rate Multiplexer .......................................................... 148
    B. Pulse Shaping ...................................................................... 153
        1. NRZ/RZ Conversion ...................................................... 153
        2. NRZ/Bipolar Conversion ................................................ 154
        3. RZ/Bipolar and RZ/NRZ Conversion .............................. 155
        4. Bipolar/NRZ Conversion ................................................ 155

III. Signal Regeneration ..................................................................... 155
    A. Clock Regeneration .............................................................. 155
        1. Preprocessing ............................................................... 156
            a. Threshold Method ................................................ 156
            b. Symmetry Method ................................................ 156
            c. Practical Example ................................................ 157
        2. PLL Design .................................................................. 158
            a. Oscillator Selection .............................................. 159
            b. Phase Detector .................................................... 159
            c. Acquisition Aids .................................................. 160
            d. Practical Implementation ...................................... 161
    B. Waveform Regeneration ....................................................... 162
        1. Equalization ................................................................. 162
            a. Circuit Concepts for Recursive and Transversal Filters ................................................................. 163
            b. Determining the Filter Parameters by Measurement ....................................................... 164
            c. Example of the Postcursor Equalizer with a Recursive Filter .................................................. 164
        2. Filtering ...................................................................... 166
            a. Circuit Implementation ......................................... 166
        3. Baseline Regeneration ................................................... 167
        4. Amplitude Regeneration ................................................ 170
            a. Ideal Comparator ................................................ 170
            b. Real Comparator .................................................. 170
            c. The Differential Amplifier as a Comparator ............ 171
            d. Emitter Degenerated Circuit as a Comparator .......... 176
            e. Integrated Comparators ......................................... 178
        5. Timing ........................................................................ 178
        6. Complete Regeneration Circuit ....................................... 180

References ...................................................................................... 181

## I. TASKS OF GENERATION AND REGENERATION

The general task of signal generation and regeneration may be defined as follows: taking into consideration the transmission characteristics and all sources of interference, signal forms must be found which allow the error rate between source and sink symbol sequence to be kept to a minimum.

An overall optimization which concerns all components as shown in Figure 1, Chapter 2 is not recommendable:

- From a theoretical point of view it would lead to an amount of dimensioning values which would, however, yield no clear design guides.
- From a computational aspect one has to take into account a great expenditure.

Therefore the usual procedure is to optimize the individual stages. In dealing with the task of signal generation and regeneration, we shall restrict ourselves here to the electronic aspect. We shall include the problems that occur with the generation of digital signals with very high bit rate. The first stage in achieving optimization concerns the transmitter and receiver filter. Due to problems connected with high-speed laser modulation, limitations are imposed on the transmission pulse shape. However, Chapter 2, Figure 2 shows that binary formats are most suitable for attenuation-limited systems where RZ and NRZ are the most important formats.

The main efforts are concentrated on the receiving end. Design of the three main components, i.e., receiver filter, clocked comparator, and clock regenerator can be undertaken separately. Clock and amplitude regeneration can be carried out independently: whereas an amplitude decision has to be made within each symbol period, this does not apply to timing, which is subjected to extremely slow fluctuations, and can be regarded as constant throughout thousands of symbol periods. For practical reasons, the receiving filter and comparator are implemented as separate components. Furthermore, it is expedient to make a distinction between equalizer and filter stages in the case of the receiving filter. The task of the equalizer circuit is to equalize the transmission channel with respect to magnitude and phase, whereas the filters serve to restrict bandwidth in order to achieve a reduction in noise. Here a compromise must be found, to avoid deterioration of the eye diagram measured at the input to the comparator. Due to the ease of implementation at high-rate systems, the equalizer and filters introduced here contain transmission line elements for determining the frequency response.

## II. SIGNAL GENERATION

### A. High-Rate Multiplexer

The multiplexing of a certain number of data streams to one data stream has been standardized for the European PCM (pulse code modulation) hierarchy up to a bit rate of 565 Mb/sec by means of the CCITT (Comité Consultatif International Télégraphique et Téléphonique) Recommendation G703. Due to the requirement imposed upon them, e.g., processing of plesiochrone data streams, these multiplexers are relatively complicated. Since integrated circuits for the gigabit per second (Gb/sec) range are not commercially available, simple multiplex methods must be chosen for bit rates > 565 Mb/sec. In this respect one important prerequisite is the synchronism of the data streams.

One simple method consists of parallel series conversion with the aid of the parallel, loadable shift register. Since integrated shift registers for the gigabit range are not yet available and implementation using discrete components is not practicable, this procedure must be ruled out for the time being.

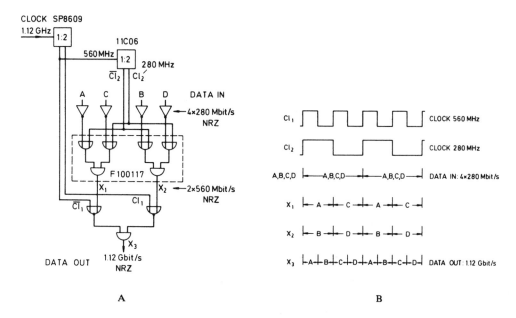

FIGURE 1. Block diagram of a 4:1 multiplexer with maximum output bit rate of 1.3 Gbit/sec. (A) Circuit diagram; (B) pulse diagram.

Cyclical sampling of the incoming data streams and the sequential arrangement of the bits produces the same data stream as that achieved with parallel series conversion. This multiplexer can be produced with relatively little effort. A number of publications contain details both about the implementation of this method using discrete components such as step recovery diodes[1,2], metal semiconductor field effect transistors (MESFETs)[3,4] bipolar transistors[5], and about implementation in the form of an integrated circuit.[6-9]

Figure 1 shows a two-stage 4:1 multiplexer using discrete components, which can be used for an output bit rate up to 1.3 Gb/sec. The first stage (dashed line) and the dividers consist of a commercially available emitter coupled logic integrated circuit (ECL-IC). The gate circuits in the second stage are built up with discrete transistors employing ECL technology.

Figure 2 shows a somewhat different circuit concept for implementing a 2:1 multiplexer for 2.24 Gb/sec using discrete components. The multiplexer consists of two sampling stages, an adder, and a limiting amplifier. The operating points of the four transistors in the sampling stage, Figure 3, are selected so that $T_2$ is switched off when the data signal is "high". $T_3$ then functions as a degenerative emitter circuit and transmits the clock signal to the output. With the data signal at "low", $T_1$ is switched off and the emitter potential of $T_3$ is shifted so by $T_2$ that $T_3$ is switched off ($u > u'$) too, resulting in a "high" at the output. Figure 4a shows the eye diagram of a sampled 1.12 Gb/sec signal. A 1.12-GHz sine signal is used as a clock signal, the amplitude of which is limited by means of the transistors $T_3$ and $T_4$, and it is converted into a signal with a rectangular signal shape. The output signals of the two samplers, Figure 2, are added and the combined signal is fed through a limiting amplifier to improve the pulse shape. Figure 4b shows the output signal of the 2.24 Gb/sec NRZ signal.

An integrated 4:1 multiplexer which can typically be used up to 1.5 Gb/sec and up to 2 Gb/sec in a modified version, is described.[8] The multiplexer was implemented using fast silicon bipolar technology (Figure 5). Depending on the bit combination at the two select inputs $S_0$ and $S_1$, one of the four data inputs $D_1$ to $D_4$ is switched to the output. If two clock signals with the same frequency and phase shift of 90° were used

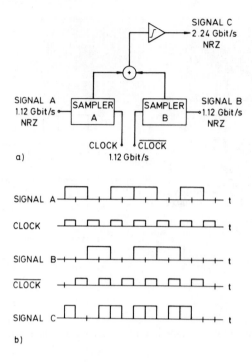

FIGURE 2.    Block diagram of a 2:1 multiplexer for an output bit rate of 2.24 Gb/sec. (A) Circuit diagram; (B) pulse diagram.

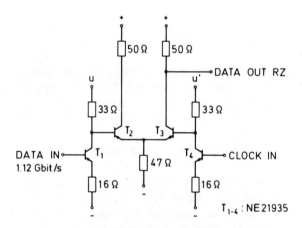

FIGURE 3.    Sampler circuit of the 2:1 multiplexer.

for the select signals, the four input signals are sampled cyclically. One major advantage in this respect is that the frequency of the select signals only needs to be one quarter of the bit frequency of the multiplexed output signal, e.g., for 1.12 Gb/sec a frequency of f = 280 MHz is required. The data inputs are ECL-compatible. The peak-to-peak output voltage is 400 mV at each output. The power dissipation amounts to $P_D$ = 220 mW at a supply voltage of 5 V. Figure 6 shows the differential output signal for 1.12 Gb/sec.

A 4:1 multiplexer for 3 Gb/sec is described.[9] It comprises three integrated 2:1 multiplexers, employing the same technology as that in the previously described multiplexers. It is a two-stage multiplexer in accordance with the principle displayed in Figure 1.

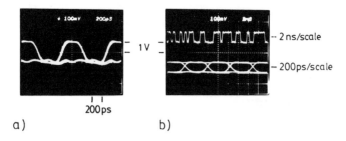

FIGURE 4.    (a) Eye diagram of the output signal (inverted) of the sampler circuit, 200 psec per division; (b) output signal of the 2:1 multiplexer for 2.24 Gb/sec; upper trace: arbitrary bit pattern, 2 nsec per division, lower trace: eye diagram, 200 psec per division.

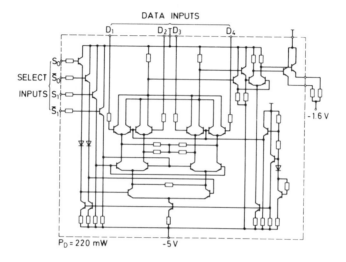

FIGURE 5.    Integrated 4:1 multiplexer[8] with maximum output bit rate of 1.5 Gb/sec.

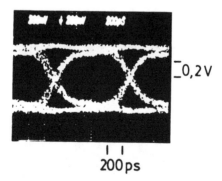

FIGURE 6.    Output signal of the integrated 4:1 multiplexer for 1.12 Gb/sec, 200 psec per division.

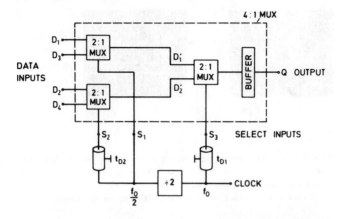

FIGURE 7.     Block diagram of the 4:1 multiplexer for 3 Gb/sec.

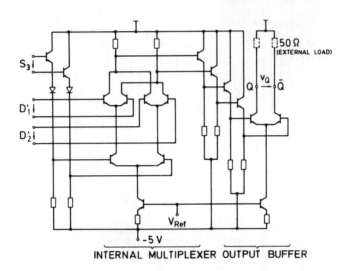

FIGURE 8.     Circuit diagram of the integrated 2:1 multiplexer.

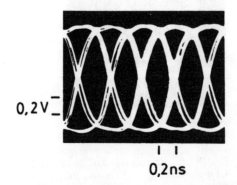

FIGURE 9.     Eye diagram of the differential output
voltage $V_Q$ of the multiplexer for 2.8 Gb/sec.

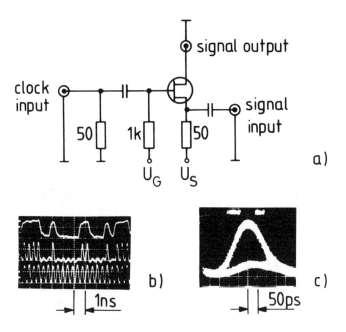

FIGURE 10.   Sampler for 2.24 Gb/sec signals with FET. (a) Circuit; (b) word pattern 2.24 Gb/sec (upper trace: NRZ input signal, middle trace: RZ output signal, lower trace: clock signal; (c) eye diagram of the output RZ signal.

Figure 7 shows the block diagram of this 4:1 multiplexer and Figure 8 the circuit diagram of the integrated 2:1 multiplexer. The differential output voltage $V_Q$ for 2.8 Gb/sec is shown in Figure 9.

## B. Pulse Shaping

Among the current systems for optical transmission the terminating equipment relies on electronic circuits processing signals in NRZ format. Nevertheless it may prove more suitable to use signals in RZ format from the viewpoint of theoretical system optimization[10] (see Chapter 2), or due to the demands imposed on the modulation behavior of lasers.[11] To avoid the difficulties associated with DC-coupled analogue and digital electronic circuits, DC-free signals can be used, e.g., in bipolar form, thereby eliminating the problems of baseline shift. NRZ, RZ, and bipolar signals occur at differing places in a transmission system. They are generated and converted from one shape to another by pulse-shaping circuits. The circuits considered here are also components of the regeneration circuits dealt with in Section III.

### 1. NRZ/RZ Conversion

For NRZ/RZ conversion a sampler circuit is required. This circuit can also be used as the timing device of the clocked comparator after Figure 1, Chapter 2 (see Section III.5). The sampling circuits for NRZ/RZ conversion have the widest bandwidth of any of the circuits discussed. According to the required pulsewidth of the sampling pulse, the fundamental and harmonics of the clock signal must be processed, so in the processing of Gb/sec signals limits are very rapidly set on any reduction in the sampling pulsewidth. In addition to sampling circuits with dual-gate field effect transistors (DuGats) with four electrodes,[12] simpler circuits can be implemented using normal field effect transistors (FET) with three electrodes.[13]

Figure 10a shows the circuit diagram for a sampler that can convert 2.24 Gb/sec

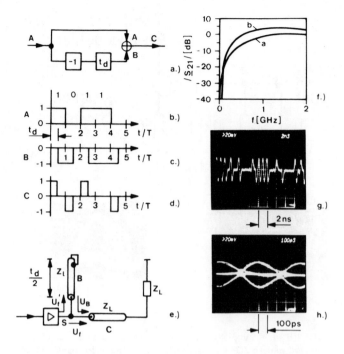

FIGURE 11.    Two-branch transversal filter for generating a bipolar signal from an NRZ signal. (a) Block diagram; (b) to (d) signal diagram; (e) basic circuit diagram of implemented circuit; (f) frequency response for conversion of 2.24 Gb/sec signal from NRZ to bipolar format (a: calculated according to [1], normalized according to 0 dB, b: measured curve of circuits s[e], [g], and [h] measured output signals).

NRZ signals into RZ signals with a 50% half-width of approximately 170 psec. The Gb/sec signal which is to be sampled enters the source, and the clock signal is applied to the gate of the FET. When the input signal at the source is high, the clock signal cannot switch on the transistor and no signal appears at the output. When the output signal at the source is low, the transistor conducts during the sample pulse period of the clock signal. An RZ signal corresponding to the clock frequency is generated at the circuit output. Figure 10b displays word patterns from the three signals that occur during sampling. Uppermost is the input signal of a 2.24-Gb/sec NRZ signal in the middle the sample RZ signal and below the sine-shaped 2.24 GHz clock signal. Figure 10c shows the 2.24-Gb/sec RZ output eye diagram with a half-width of 170 psec.

### 2. NRZ/Bipolar Conversion

In this context the term "bipolar signal" refers to a signal that only transmits the change in slope of a NRZ signal. At the leading edge of the signal a positive pulse is produced, and the declining signal slope produces a negative pulse (differentiation of an NRZ signal). Between the signal an 0 is read.

Basically, the circuit can be set up as in Figure 11a. It consists of a dual-branch transversal filter in which, following branching, a signal inversion and signal delay $t_d$ is introduced in one branch. Following summation, the bipolar signal is created as shown in the time curve diagrams 11b to d. The frequency response of this filter is as follows:

$$\underline{S}_{BIP} = 1 - e^{-j\omega t_d} \tag{1}$$

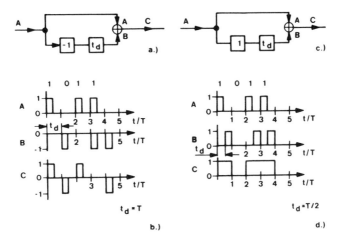

FIGURE 12. Two-branch transversal filter for format conversion. (left) RZ/bipolar conversion; (right) RZ/NRZ conversion ([a] and [c] block diagrams, [b] and [d] signal diagrams).

This response can be implemented by the circuit of Figure 11e. The NRZ signal to be converted is delivered to an amplifier with a high-impedance output. At point S the signal is branched between the outgoing line C (signal $U_f$) which is reflection-free terminated into the resistor $Z_L$ and the short-circuited stub line B with a delay time of $T_d/2$ (signal $U'_f$). Both lines C and B have a characteristic impedance of $Z_L$. The signal fed into the stub line is reflected at the short circuit and returns with the opposite sign ($U_B$) to point S, where it appears after the total delay time $t_d = 2t_d/2$. Due to the high-impedance amplifier output, it enters the outgoing circuit C without reflection and is superimposed on the signal $U_f$. Figure 11f shows calculated and measured amplitude responses for an implemented circuit; 11g and h show the word pattern and eye diagrams for a 2.24-Gb/sec bipolar signal.

### 3. RZ/Bipolar and RZ/NRZ Conversion

Two-branch transversal filters as described above can also be used for RZ/bipolar conversion and for RZ/NRZ conversion. In accordance with the details given in Section II.B.2, only the line length of the stub line and its terminating resistor need to be altered. Block diagrams for the appropriate transversal filters, together with the time diagrams, are given in Figure 12.

### 4. Bipolar/NRZ Conversion

When converting from bipolar to NRZ signals, the missing low-frequency components in a bipolar signal must be generated. This can only be achieved using nonlinear circuits.

Figure 13 shows a Schmitt trigger for bipolar/NRZ conversion. Use is made of the hysteresis properties of the circuit (Figure 13b). The positive slope of the bipolar signal sets the Schmitt trigger in the high state, and the negative switches it back to the low state. An example of an application is given in Section III.B.5.

## III. SIGNAL REGENERATION

### A. Clock Regeneration

If the timing is to be regenerated without any knowledge of the information content, a process must be used that generates the bit frequency component independently of

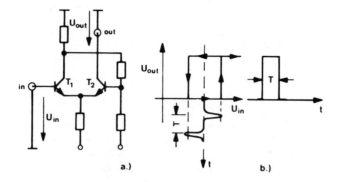

FIGURE 13.     Schmitt trigger for conversion of bipolar to NRZ signals. (a) Basic circuit diagram; (b) signal diagram.

the bit pattern. This takes place in the so-called preprocessing unit (Chapter 2, Figure 1). It is essential that the received signal contains sufficient timing information and that the phase of the spectral line produced remains constant. The bit frequency component is not directly available in the case of highly utilized communication systems,[14] and must be generated by nonlinear means. If the signal contains a spectral bit frequency line of sufficient amplitude, no preprocessing has to be carried out at all.

Signals obtained by preprocessing still contain some unwanted frequency components, and these must be suppressed. There are basically two ways of doing this: a passive method, consisting of an oscillating circuit to reduce noise effects followed by a hard limiter to eliminate amplitude variations,[15] and an active method with phase-locked loop (PLL), which contains a local oscillator.

On the theoretical side there is no significant difference between the passive solution and a first-order PLL. A reduction of the timing information combined with insufficient stability of the components leads to increased phase fluctuations. In the case of second-order PLL this interrelationship can be reduced by proper design of the loop filter to any desired amount. On the other hand, this does adversely affect the pull-in behavior. It is therefore necessary to make provision for an acquisition aid (Section III.A.2.c). A PLL equipped in this way remains unaffected by fluctuations of the clock content.

## 1. Preprocessing

### a. Threshold Method

The incoming signal is filtered in such a way that the overall impulse response — concerning the signal path from the laser/modulator-transmitter filter up to the matching filter output — produces a pulse that satisfies the second Nyquist criterion[16] (see Figure 14). This implies that the eye diagram occupies the symbol width. A comparator set to the appropriate threshold produces a signal that is then both differentiated and rectified. The output pulse train contains a bit frequency component. Following this solution all demands must be met with a high degree of precision in order to avoid jitter components. This is immediately understandable considering the influence of the comparator threshold.

### b. Symmetry Method

The incoming signal is filtered in such a way that the overall impulse response produces a symmetrical band-pass signal with half the bit frequency as center frequency[17] (see Figure 15). This signal is then treated nonlinear, e.g., squared. After high-pass filtering, the resulting signal contains zero-crossing with the same spacing as the sym-

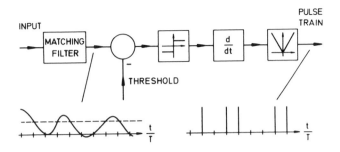

FIGURE 14.    Diagram for threshold preprocessing.

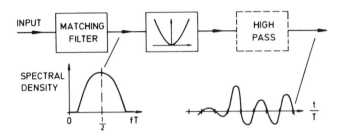

FIGURE 15.    Diagram for symmetry preprocessing.

bol durations. This filtering is only mentioned here for reasons of completeness; it is carried out by the following band-pass filtering. The advantage of this method is that only limited demands are imposed on the prefiltering process. One disadvantage is that the nonlinear process must be carried out with symmetry, otherwise the phase will be dependent on the bit pattern. Monolithic integrated circuits do not generally satisfy these requirements.

### c. Practical Example

In the Gb/sec range the symmetry method can easily be implemented using delay-line clipping. The example shown in Figure 16 is suitable for transmission in the NRZ format with Gaussian channel pulse response. The delay-line configuration has a frequency response as follows:

$$G(j\omega) = 1 - \exp(-j\omega t_d) \tag{2}$$

The electronic full wave rectifier acts as a nonlinear device. In order to fulfill the high requirements at the symmetry, the structure is a symmetrical one. It consists of a unity gain phase splitter and two half-wave rectifiers, which have been realized as zero-biased emitter degenerated circuits. The output signal contains a bit frequency component with data-independent phase. Because the product of $G(j\omega)$ with the frequency response of an NRZ-Gaussian pulse cannot yield a symmetrical shape, the condition that a symmetrical band-pass signal should be available after prefiltering is not ideally fulfilled, but in practical terms in a sufficient manner. In case of other signal formats or other channel pulse responses the use of more complex filter configurations may be necessary, e.g., a second delay-line stage. Figure 17 shows the amplitude of the generated bit line. Obviously the dimensioning of the delay time $t_d$ is not critical. The optimum $t_d$ varies depending on the channel bandwidth between T/2 and T, T being the symbol period. The lower value is obtained for bandwidth $f_{bw} \rightarrow \infty$, i.e., rectangular

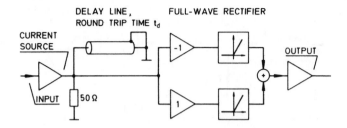

FIGURE 16.    Implementation of symmetry method.

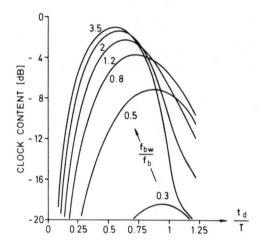

FIGURE 17.    Clock content for NRZ format via Gaussian channel. Parameter is $f_{bw}/f_b$ (3 dB bandwidth per bit frequency).

waveshape, the upper being obtained for $f_{bw} \to 0$, however, in this case no amplitude regeneration is possible because the eye diagram is closed.

## 2. PLL Design

Figure 18 shows the basic scheme of a PLL. The main parameters of the compounds determining performance are

- $K_D$ — the gain of the linearized phase detector
- $F(j\omega)$ — the frequency response of the loop filter
- $K_o$ — the frequency conversion factor of the voltage-controlled oscillator (VCO)

Given these values, the essential design parameters of a second order PLL, the natural frequency $\omega_n$, and the damping factor $\zeta$ can be readily computed. (For details see Gardner.[18]) If no special measures are undertaken, $\omega_n$ and $\zeta$ are dependent upon the bit pattern. The design must be carried out assuming the worst case condition, i.e., for that bit pattern containing the minimum amount of timing information. In order to avoid extensive jitter accumulation, a high damping factor should be selected when a long haul link contains a great number of amplifiers, e.g., $\zeta = 5.$[17] For reasons of noise reduction $\omega_n$ should be dimensioned as small as possible. However, a low $\omega_n$ leads to a high pull-in time and to high requirements on the VCO so that a compromise must be found.

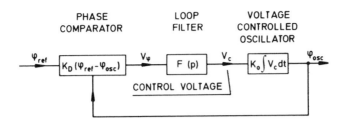

PHASE COMPARATOR     LOOP FILTER     VOLTAGE CONTROLLED OSCILLATOR

FIGURE 18.    Basic scheme of a PLL.

## a. Oscillator Selection

PLL circuits often do not reach the calculated performance because insufficient attention was paid to the question of oscillator quality. A few fundamental considerations should be given to control theory aspects. As determined by the hardware, the PLL design is governed by only one degree of freedom, i.e., the loop filter characteristic. More complex structures lead to an expense which is much higher than that of the passive solution. The most important quantity in this relation is the noise bandwidth $B_L$. For a second order PLL, $B_L$ depends only on $\zeta$ and $\omega_n$. It holds in principle that phase fluctuations of the oscillator are only suppressed within the noise bandwidth $B_L$ of the PLL. Therefore it makes no sense to design $B_L$ to values beyond the noise bandwidth of the oscillator. For the case that the noise reduction of the reference signal is insufficient, the only solution is to increase the quality of the oscillator. No doubt the phase slope of the feedback path is a major influencing factor. Considering this, the following order can be established for the gigahertz range:

- RC — oscillators
- LC — oscillators
- Transmission line oscillators[19]
- Quartz oscillators

Aberrations resulting from long-term drift and temperature behavior must be rated to the tuning range. It is rather costly and complicated to set up quartz circuits for the extremely high overtones required, which must also be tunable. The transmission line VCO can be achieved in a very compact form in the gigahertz range, with very high quality, and in addition no particular problems are encountered with tuning. Since it can also be set up as a thick film circuit, it provides the most favorable solution. With respect to LC and RC oscillators, it must be taken into account that the frequency-determining components L, C, or R are generally deteriorated by parasitic impedances as described in Chapter 7.

## b. Phase Detector

In those systems with medium or low signal-to-noise ratios (SNRs) multipliers are used almost exclusively as phase meters. The implementation of the multiplication by active devices has to be ruled out at present due to the problem of offset drift. A feasible solution is the use of a double-balanced mixer. However, since such a mixer contains two transformers, monolithic integration is hardly possible.

For higher SNR, i.e., for systems under consideration here, digital phase meters may be used. It must be pointed out that digital phase meters do not evaluate the whole waveform but trigger out one instant from the signal. These components may be well suited for integration. However, there are problems which must be considered carefully:

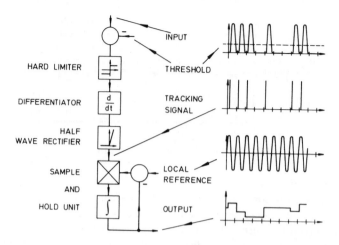

FIGURE 19.    Diagram and timing of sample and hold phase detector.

- Tolerance of the high/low level voltages may lead to offset drift effects
- The gate propagation delay and other speed limitations may deteriorate the detector characteristic

One interesting concept for the gigahertz range is provided by the sample and hold phase detector,[20] which operates discrete in time but analogue in amplitude. Figure 19 shows a basic circuit diagram. Preprocessing is carried out using the threshold method, with only the postive transitions being evaluated. Each time the input signal exceeds the threshold, the output is set to the momentary voltage of the local reference. The waveform of the local reference is identical with the resulting phase meter characteristic, i.e., a sinusoidal oscillator voltage leads to a sinusoidal characteristic. The sample and hold unit poses the main problem when implementing this system. The outstanding feature is that no positive transition occurs if zero is transmitted and the final value is maintained; the conversion factor of the detector is therefore not dependent on the bit pattern.

### c. Acquisition Aids

Using a PI filter (also referred to as a bridged integrator) as the loop filter, the residual error becomes zero; however, an acquisition aid must be provided in this case. Numerous suggestions have been made in this respect.[18] Frequently an unlock condition is derived using a second phase meter.

For a given unlock condition the tuning range is passed continuously, possibly with different speeds, searching for signals onto which it can be locked. This method is referred to as sweeping. Long lock-in times can occur, regardless of whether the initial frequency error was small or large.

Thus the application of an additional frequency control loop may be desirable. For this a sign-dependent frequency difference meter for the gigahertz range is required. There are several frequency-sensitive integrated phase meter circuits available — however, with insufficient speed. A solution to this problem is provided by beat frequency control.[21] Figure 20 shows the block diagram. A second phase meter is fed via a delay line which causes a phase delay of approximately 90°. During the unlocked PLL condition, the output voltage of this PLL phase meter is sampled at the instant at which the output voltage of the auxiliary phase meter crosses zero with positive slope. The number of zero crossings per unit of time is proportional to the absolute value $|\Delta f|$ of

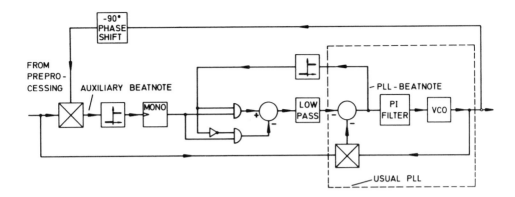

FIGURE 20.    Beat frequency control schematic.

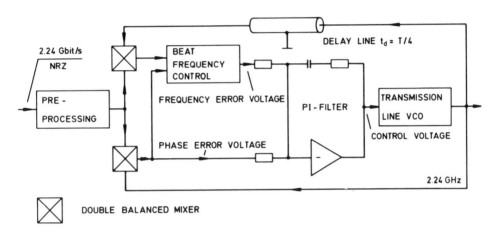

FIGURE 21.    Schematic of a complete regeneration circuit.

the beat note frequency. Thus the DC-component of the output voltage of the mono-flop is a measure for this value. The sign of Δf is derived from the sign of the PLL beat note at the sampling instant. By averaging the signed samples, a voltage is obtained proportional to the frequency difference Δf between VCO and bit frequency. This voltage is used for frequency control. The main advantage is that the only high-frequency component is the auxiliary phase meter, whereas the highest frequency that occurs at the other components is determined by the width of the tuning range. An acquisition range equal to the tuning range of the VCO can be obtained by proper dimensioning.

### d. Practical Implementation

In the following, some features of a complete regeneration circuit PLL are given which has been developed for clock extraction from a 2.24 Gb/sec NRZ data stream. Figure 21 shows the scheme of the circuit:

- Due to the NRZ signal format, preprocessing must be applied: the prefiltering is carried out as shown in Figure 16.
- The 2.24-GHz VCO is of the varactor-tuned transmission line type and was implemented using thick film technology. Its tuning range is ± 20 MHz and the temperature coefficient amounts to − 200 kHz/K.
- The phase meter is a commercially available, double-balanced mixer module.

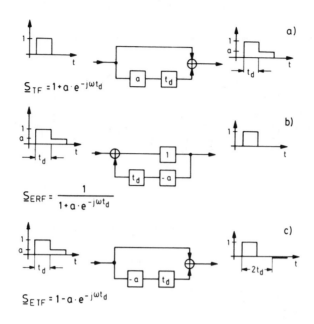

FIGURE 22.    Postcursor generation and equalization. (a) Transversal filter for generation of postcursor; (b) recursive filter for complete elimination of postcursor; (c) transversal filter for reduction of postcursor.

- The loop filter is of the PI type. Thus an acquisition aid is indispensible.
- For acquisition, the beat note frequency control method has been used as shown in Figure 20, the auxiliary phase meter being of the same type as the PLL phase meter. This causes the acquisition range of the PLL to become equal to the tuning range of the VCO. The maximum time required to achieve lock-in is 1.7 msec.
- When the mean transition density of the NRZ input signal is 50% of the maximum transition density, the noise bandwidth of the PLL is 0.96 MHz and the damping factor is 2.7. The rms output jitter is 3°, and is due mainly to instability of the modulated laser transmitter.

## B. Waveform Regeneration

### 1. Equalization

Although in the optical long wavelength region the transmission properties of the fiber itself may be regarded as ideal, impairments of the electronic and electro-optical devices installed in Gb/sec systems can lead to signal distortions which should be cancelled prior to the amplitude and time regeneration circuit in the receiver. The phase and amplitude distortion in the transmission system resulting from these effects can be eliminated by an equalization circuit at the receiving end. If the tailing effects of photodiodes are to be considered, there is a problem involving the equalization of nonsymmetrical pulses with positive postcursors. Recursive and transversal filters will be described here, capable of postcursor equalization.[22][23]

As is generally known, a signal with positive postcursor is obtained at the output of a two-branch transversal filter as shown in Figure 22. Its frequency response is

$$\underline{S}_{TF} = 1 + a\,e^{-j\omega t_d} \tag{3}$$

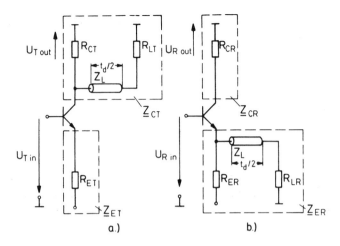

FIGURE 23. Basic circuit diagrams of (a) transversal filter; (b) recursive filter.

a is the postcursor amplitude and $t_d$ is the delay time by which the postcursor appears after the main signal. The signal can be ideally equalized using the recursive filter of Figure 22b with the transfer function

$$\underline{S}_{ERF} = \frac{1}{\underline{S}_{TF}} = \frac{1}{1 + a\,e^{-j\omega t_d}} \tag{4}$$

Approximate equalization can be achieved by using a two-branch transversal filter as shown in Figure 22c with the frequency response

$$\underline{S}_{ETF} = 1 - a\,e^{-j\omega t_d} \tag{5}$$

A residual postcursor is left with a reduced amplitude. In the case of postcursors with small amplitudes, the frequency responses of Equations 4 and 5 are roughly the same as for $a \ll 1$.

$$\underline{S}_{ERF} = \frac{1}{1 + a\,e^{-j\omega t_d}} \approx 1 - a\,e^{-j\omega t_d} = \underline{S}_{ETF} \tag{6}$$

#### a. Circuit Concepts for Recursive and Transversal Filters

Figure 23 shows the basic circuits for transversal and recursive filters which have frequency responses in accordance with Equations 4 and 5. In both cases current feedback circuits in common emitter circuits are shown. For the voltage gain V the following applies:

$$\underline{V} = \frac{\underline{U}_{out}}{\underline{U}} = \frac{\underline{Z}_C}{\underline{Z}} \tag{7}$$

$\underline{Z}_C$ and $\underline{Z}_E$ are the complex collector and emitter impedances. In the circuits described here the collector impedance of the transversal filter $\underline{Z}_{CT}$ and the emitter impedance of the recursive filter $\underline{Z}_{ER}$ have the same configuration. They consist of a resistor $R_{CT}$ or $R_{ER}$ located at the collector or emitter, respectively, a transmission line with characteristic impedance $Z_L$, and a delay time of $t_d/2$ and a terminating resistor $R_{LT,R}$ determining the real reflection factor at the end of the line. The emitter impedance $R_{ET}$ of the transversal filter and the collector impedance $R_{CR}$ in the case of the recursive filter are also real. If for $R_{CT}$ and $R_{CR}$ the value of the characteristic line impedance $Z_L$ is chosen, it can be easily proved that the voltage gain of the transversal filter $\underline{V}_{TF}$ is

$$\underline{V}_{TF} = 1 + b\,e^{-j\omega t_d} \tag{8}$$

and for the recursive filter

$$\underline{V}_{RF} = \frac{1}{1 + b\,e^{-j\omega t_d}} \tag{9}$$

Without going into details, it should be pointed out that the relationships given here can only be applied exactly to ideal transistors, e.g., whereas ideally there should be a high-impedance output in the case of the transversal filter described (see Section II.B.2), in the case of the recursive filter there should be a low impedance seen from the stub line into the emitter.

The b values (8) and (9) correspond to the respective reflection factors at the end of the line, and are determined by the values for the terminating impedances $R_{LT}$ or $R_{LR}$. $t_d/2$ represents the delay time of the transmission line. Equations 9 and 8 correspond exactly to the equations for the recursive and transversal filters in Equations 4 and 3 or 5. In the case of the transversal filter, b is substituted as follows: b = −a (see Figure 22b), which is achieved by an appropriate value for $R_{LT}$. If the inputs and outputs of the circuits are terminated reflection-free, the voltage gain V and the transmission factor $\underline{S}_{21}$ of the circuits which can be measured with a network analyzer are identical.

### b. Determining the Filter Parameters by Measurement

When the frequency response of the transmission system is not given analytically, filter simulation with adjustable parameters a and $t_d$ can be carried out easily in the case of the two-branch transversal filters for postcursor equalization as shown in Figure 22c. The function of amplitude gain a, delay $t_d$, and the summation can be simulated by a dual-trace oscilloscope possessing the internal functions VARIABLE GAIN, INVERT, DELAY, and ADD. A simple power divider can be used to split the signal at the filter input. In Figure 24, a 1.12 Gb/sec RZ signal with jitter and pulse interference is displayed in order to demonstrate the simulation process.[24] Figure 24b shows the two signal streams of the simulated filter branches, which produce the equalized signal (Figure 24c) following addition. The oscilloscope functions VARIABLE GAIN and DELAY were adjusted until an optimum eye opening for the sum of the two signals was achieved.

The values for a and $t_d$ can be obtained directly from Figure 24b in the example a = −0.27; $t_d$ = 0.7 nsec. For the corresponding recursive filter the values are virtually identical (except for the reversed sign of a) in the case of small postcursor amplitudes. In the case of higher postcursor amplitudes, optimization for the recursive filter must take place directly on the equalizer circuit, with the assistance of a phase shifter, an adjustable attenuator, and a sliding short.

### c. Example of the Postcursor Equalizer with a Recursive Filter

Figure 25 shows the circuit diagram for a recursive filter.[25] The open-ended line $L_1$ on the collector corrects the frequency response of the nonideal transistor. The resistive input network 33 and 18 Ω was chosen to minimize the input reflection coefficient. The behavior of the circuit in the time and frequency domain was investigated using an analysis program as described in Chapter 7.

An input signal with a postcursor was synthesized by calculation (Figure 26a, curve a). The curve b displays the equalized signal at the output of the circuit. In Figure 26 c

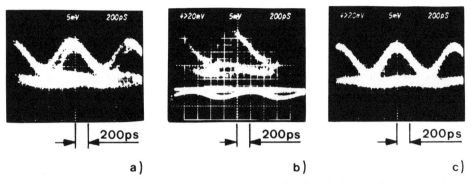

FIGURE 24. Postcursor equalization of a 1.12 Gb/sec RZ signal received over a distance of 21 km — filter simulation on an oscilloscope. (a) Received signal; (b) received signal and compensating signal; (c) addition of received signal and compensating signal.

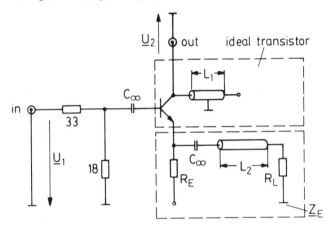

FIGURE 25. Complete circuit diagram of recursive filter for postcursor equalization.

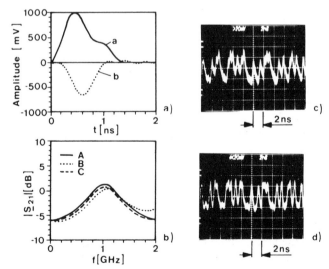

FIGURE 26. Behavior of recursive filter for postcursor equalization in the time and frequency domain. (a) Computed synthesized pulse with postcursor at the input of the filter (a) and equalized pulse (b); (b) amplitude response of transmission factor $\underline{S}_{21}$ (A: calculated according to (7), B: measured, C: calculated with an analysis program); (c) 2.24 Gb/sec NRZ signal after transmission over a distance of 21 km; (d) equalized signal.

and d signal traces of real received and equalized 2.24 Gb/sec signals are shown. Figure 26b depicts the amplitude of the transmission factor $\underline{S}_{21}$ of the circuits. Curve A shows the calculated one according to Equation 9, taking into account the input voltage divider. B and C show them measured and calculated with the analysis program. The impedance $R_L$ which terminates the transmission line of Figure 25 amounted to 115 Ω equivalent to a reflection coefficient factor of 40%. The delay time $t_d/2$ of the line was 220 psec.

### 2. Filtering

In this context the term "filter circuit" is used for those circuits which reduce the high-frequency noise component following channel equalization. If colored noise occurs, certain frequency components can be selectively tuned out.[26] It is not the intention here to discuss in detail the conventional methods of filtering such as Gaussian, Bessel, and maximum flat filtering, which are sufficiently well documented, and for which possible solutions exist already for processing signals in the Gb/sec range.[27-28] Here the possibilities for filtering and bandwidth reduction with easy to implement transversal filters are demonstrated. For amplitude and phase characteristics of these filters relevant literature is mentioned.[22-23]

The example chosen is a four-branch transversal filter as shown in Figure 27a, which provides low-pass limitation of an NRZ signal. This filter has a periodic frequency response. Use is only made of that frequency range in which the signals have considerable spectral components. Additional filters should be used to suppress the higher-frequency components. The frequency response of the filter is as follows:

$$\underline{S}_4 = 1 + e^{-j\omega t_{d1}} + e^{-j\omega t_{d2}} + e^{-j\omega t_{d3}} \tag{10}$$

The time curves in diagrammatic form are shown in Figure 27b for $t_{d3} = t_{d1} + t_{d2}$. The signal E clearly shows the pulse broadening based on the resulting low-pass effect.

### a. Circuit Implementation

In the processing of Gbit/sec signals, the splitting and addition of the signals as shown in Figure 27a produces considerable difficulties in the implementation of such circuits. An arrangement such as that shown in Figure 27c leads to the same result. It consists of two two-branch transversal filters connected in series. The time curves in diagrammatic form are shown in Figure 27d. The frequency response of the filter is

$$\underline{S}_4' = (1 + e^{-j\omega t_{d1}})(1 + e^{-j\omega t_{d2}}) \tag{11}$$

$$= 1 + e^{-j\omega t_{d1}} + e^{-j\omega t_{d2}} + e^{-j\omega(t_{d1} + t_{d2})} \tag{12}$$

For $t_{d1} + t_{d2} = t_{d3}$ it is identical with Equation 10. The amplitude response is shown in Figure 27e. The maximum value at $f = 0$ was normalized in this case to 0 dB. The delay time $t_{d1}$ corresponds to a quarter of the clock period of a 1.12 Gb/sec signal. $t_{d1} = 1/4T = 1/4 \cdot 1.12$ nsec = 223 psec. $t_{d2}$ is equivalent to half the timing period $t_{d2} = 1/2T = 446$ psec.

For comparison in Figure 27e, the amplitude responses of individual stages in accordance with Figure 27c are also given.

The behavior in the time domain of a realized two-stage transversal filter as in Figure 27c is depicted in Figure 28. Eye diagrams and signal sequences of a 1.12 Gb/sec pseudorandom NRZ signal are shown in Figure 28 a and b. In Figure c and d the influence of the first filter stage is demonstrated. The staircase structure of the signal is quite pronounced. Figure 28e and f shows the outgoing signal for the complete filter. While

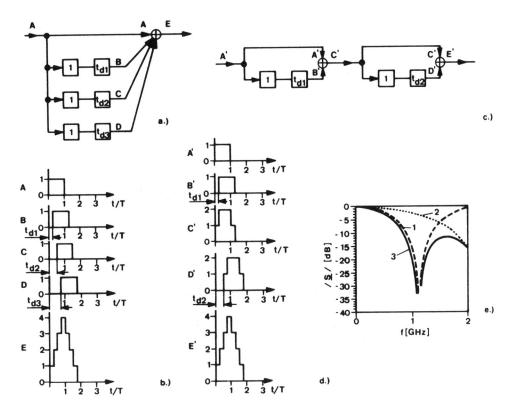

**FIGURE 27.** Transversal filter for low-pass limitation. (a) Block circuit diagram of a four-branch transversal filter; (b) diagram of time curves for (a); (c) block circuit diagram of equivalent two-stage transversal filter; (d) diagram of time curves for (c); (e) amplitude responses of filter (c) for 1.12 Gb/sec signal (1: filter with cos-shaped response, $f_0 = 1.12$ GHz; 2: filter with cos-shaped response, $f_0 = 2.24$ GHz; 3: both filters connected in series).

the eye aperture is still completely open at the instant of sampling in the middle of the signal, outside this instant the horizontal and vertical eye aperture is considerably reduced. Whereas the input signals have edges < 200 psec the edges of the ouput pulses are > 800 psec.

In order to suppress additional frequency components, additional nulls of the filter frequency response can be introduced by cascading transversal filters. For example, a partial response signal[29] is easily generated by introducing a third stage of a two-branch transversal filter with an additional null at the half clock frequency.

### 3. Baseline Regeneration

When signals with low-frequency and DC-spectral components are transmitted in a high-pass system — e.g., capacitively coupled stages — baseline wander occurs. The baseline shift depends on the momentary duty cycle of the signal. When baseline restoration is desired it has to be performed prior to the comparator in the receiver. Only nonlinear processing is suitable to abolish the spectral null at DC.

Clamping circuits are only of limited use for baseline regeneration. Basically they do not eliminate the pulse tilt of a high-pass signal. Due to the nonideal reverse impedance of the clamping diode and the finite input impedance of the isolating amplifier, it is only possible to achieve a reduction in the baseline shift. Complete recovery of the DC-component is not possible.

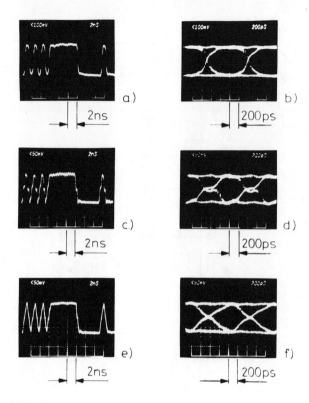

**FIGURE 28.**     Examples of impulse shaping for bandwidth reduction using transversal filters. (a) and (b) Input signal: 1.12 Gb/sec NRZ; (c) and (d) output signal of a transversal filter with cos-shaped frequency response ($f_{01} = 1.12$ GHz); (e) and (f) output signal of a two-stage transversal filter ($f_{01} = 1.12$ GHz, $f_{02} = 2.24$ GHz).

As known from coaxial cable technology, the quantized feedback equalization (QFE) for DC restoration circumvents these problems. It can also be used for optical transmission systems.[30] Figure 29 shows the basic diagram for such a circuit. The high pass at the input of the circuit with a cut-off frequency $f_{gHP}$ (point 1) introduces pulse tilt and base line shift (point 2). Assuming that the circuit does not present any internal delay time, the leading edge of the incoming pulse will switch the circuit over to the high state when the threshold (point 3) is exceeded. A part of the signal is branched off from the output and fed to a low pass with the frequency $f_{gLP} = f_{gHP}$. The low pass is complementary to the high pass at the input. The complementary low-pass signal delivers the instantaneous threshold signal (point 3) of the comparator consisting of $T_1$ and $T_2$. It follows the slope of the high-pass signal. In the case of real circuits considerations should be given to the internal delay time that can be permitted. When delays in the circuit are anticipated, which lead to a late threshold correcting signal, linear predistortion of the input signal is possible — e.g., a transversal filter — which allows correction to later times.[31]

In the case of a real transmission system, the high-pass structure is of course more complicated than that assumed in the basic circuit diagram of Figure 29. The stages altogether contribute to a higher order high-pass filter. Exact equalization is not necessary if another high pass with a much higher cut-off frequency is introduced.[32] It can be shown that then the influence of the baseline wander of the other high passes may be ignored. For that reason it is sufficient to compensate for the influence of the additional introduced high pass only.

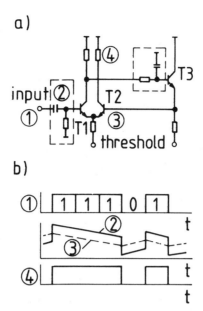

a)

b)

FIGURE 29. Circuit for baseline restoration using QFE. (a) Basic circuit diagram; (b) diagram of time curves.

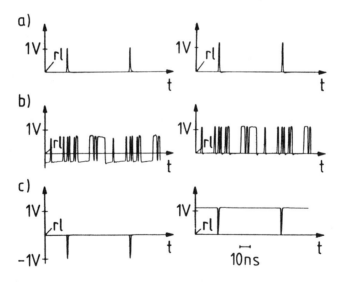

FIGURE 30. Selected word pattern to show the properties of the QFE. (left side) Signals behind high pass; (right side) signals behind decision circuit ([a] low duty cycle 1:63 high to low, [b] arbitrarily chosen bit pattern, [c] high duty cycle 63:1 high to low). rl = reference level.

Figure 30 gives examples of baseline regeneration of a 1.12 Gb/sec NRZ signal.[30] The additional high pass had a threshold frequency of approximately 10 MHz. Signals with baseline shifts prior to the additionally inserted high pass on the circuit input (left) and after the baseline regeneration at the output of the circuit for quantized feedback (right) are shown for a variety of different duty cycles. This procedure also functions

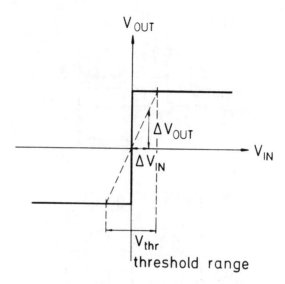

FIGURE 31.    Transfer characteristic for an ideal comparator (solid line) and for a real comparator (broken line).

correctly in the case of periodic signals with duty cycles 1:63 (above) and 63:1 (below)
high-to-low signals.

### 4. Amplitude Regeneration

The amplitude and timing regeneration are represented in Figure 1 of Chapter 2 by
the block called "clocked comparator". In this and the following section, this block
will be split up and the amplitude and timing regeneration will be considered separately.

In an amplitude regeneration circuit a decision is made as to whether the momentary
value of the amplitude of the received signal is above or below the reference level, i.e.,
the threshold. The output of the amplitude regenerator (comparator) is switched to
"high" or "low" level. This provides a binary signal at the output of the comparator.
Basically the comparator is a limiting amplifier with a very high gain. (Ideally the gain
should be infinite.)

### a. Ideal Comparator

The main features of an ideal comparator are given below:

1.    Figure 31 shows the transfer function for an ideal comparator (solid line). For
      $V_{IN} = 0$ there is noncontinuous transition between the "low" and "high" levels.
2.    The operational speed of an ideal comparator is unlimited, i.e., a "high" to
      "low" decision takes an infinitely short time and the comparator is able to make
      a subsequent decision immediately.
3.    The properties of an ideal comparator are not dependent on the amplitude of the
      input signal.
4.    The properties of an ideal comparator are not dependent on other parameters
      such as temperature or supply voltage drift, etc.

### b. Real Comparator

For $V_{IN} = 0$ the real comparator has only a finite gain (broken line, Figure 31). The
gain depends on the design of the comparator and the properties of the components

used. In the threshold range, (Figure 31) no clear decision is reached with the result that the vertical eye aperture of the input signal is reduced by this range. This in turn means that the SNR of the input signal at a given error rate must inevitably be greater in the case of the real comparator than for the ideal comparator. We shall now consider the amount by which the SNR should be increased. A worst case situation is assumed where only incorrect decisions are made in the threshold range.

Assuming an ideal comparator and a given error rate, a SNR of

$$SNR^* = 10 \log \frac{P_S^*}{P_N}$$

is required ($P_S^*$ = signal power, $P_N$ = noise power).

Given a real comparator with a certain threshold range and the same given error rate, a higher SNR of

$$SNR = 10 \log \frac{P_S}{P_N}$$

is necessary ($P_S \geqslant P_S^*$; see Figure 32). The signal-dependent noise in optical systems will not be taken into account in this consideration.

With

$$P_S^* = \frac{V_S^{*2}}{R} \quad \text{and} \quad P_S = \frac{V_S^2}{R} \qquad (R = \text{unit resistance})$$

and the difference A = SNR − SNR*, we obtain

$$A = 20 \log \frac{V_S}{V_S^*}$$

With $V_S^* = V_s - V_{thr}$ ($V_{thr}$ = voltage amplitude of the threshold range) we obtain

$$A = 20 \log \left( 1 - \frac{V_{thr}}{V_S} \right)$$

(see Figure 33).

Thus for an input signal with a peak-to-peak voltage of $V_{spp}$ = 1 V and a comparator with a threshold range of 0.1 V, we obtain a penalty of 0.9 dB. This shows that with a relatively large threshold range in relation to the peak-to-peak input voltage, a penalty must be paid which is, in general, acceptable. This result is of decisive importance in the case of high bit rates (> 500 Mb/sec) because given the components available at the present time, the implementation of a comparator with a small threshold is bound to incur a great deal of effort.

### c. The Differential Amplifier as a Comparator

In low-rate systems the differential amplifier is the most commonly used basic circuit for setting up an amplitude comparator (Figure 34). Among the advantages it offers are good temperature compensation within the circuit and the possibility for shifting the threshold via the second input.

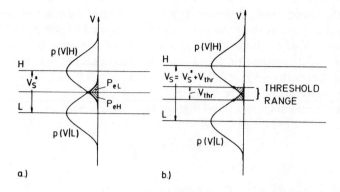

FIGURE 32.    Probability density functions p(V|L) and ;(V|H) of the interference superimposed on a binary signal $P_{eL}$ and $P_{eH}$ probability of error for the "low" and "high" levels, respectively (H = "high" level, L = "low" level). (a) Comparator with ideal threshold; (b) comparator with threshold range.

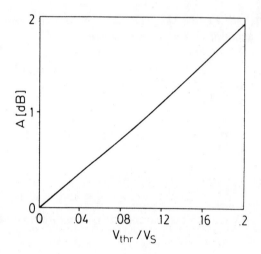

FIGURE 33.    Penalty A which has to be paid by the threshold range $V_{thr}$ of a real comparator.

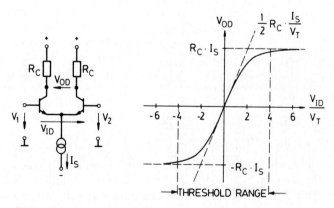

FIGURE 34.    Transfer characteristic for an ideal differential amplifier.

Figure 34 shows the static transfer characteristic for an ideal differential amplifier. The differential output voltage $V_{OD}$ as a function of the differential input voltage $V_{ID}$ is given by

$$V_{OD} = R_C I_S \tanh \frac{V_{ID}}{2V_T}$$

(thermal voltage $V_T = 25.9$ mV at 300°K).

The differential gain $\dfrac{dV_{OD}}{dV_{ID}}$ for $V_{ID} = 0$ is

$$\left. \frac{dV_{OD}}{dV_{ID}} \right|_{V_{ID}=0} = \frac{R_C \, I_S}{2V_T}$$

With $\dfrac{I_s}{V_1} = g_m$ (transconductance) we obtain:

$$\left. \frac{dV_{OD}}{dV_{ID}} \right|_{V_{ID}=0} = \frac{1}{2} R_C \, g_m$$

An input voltage of approximately 200 mV is necessary in order to drive the differential amplifier from one limitation to the other

$$V_{IDpp} \simeq 8V_T \simeq 200 \text{ mV},$$

i.e., using the differential amplifier as a single-stage comparator, a threshold range of approximately 200 mV is inherent.

So far, our considerations have applied to the quasi-static operation of a differential amplifier. We shall now turn our attention to a number of problems that arise in connection with currently available transistors (transit frequency $f_t < 10$ GHz) in the use of a differential amplifier as a comparator at high bit rates ($> 500$ Mb/sec).

For the differential amplifier shown in Figure 35a, the frequency response is depicted in Figure 36, curve 1. The frequency response was calculated for nonpacked transistors with a transit frequency of $f_T = 5$ GHz. For details of transistor modeling, see Chapter 7.

The $-3$-dB cut-off frequency amounts to $f_{bw} = 250$ MHz. As will be shown below, it is far too low for the processing of Gb/sec signals. At 500 MHz (Nyquist frequency for 1 Gb/sec) the gain will have dropped by 6.5 dB in comparison to $f \rightarrow 0$, corresponding to a reduction in the slope of the transfer characteristic in figure 34 at this frequency. The reduction in the slope results in an increase in the threshold range of 200 mV in the quasi-static case to approximately 450 mV at 1 Gb/sec. Figure 37 shows the time curves of the differential output signal for various peak-to-peak voltages of the input signal for 1 Gb/sec. For simulation, a ... 001010 ... NRZ pulse sequence with sinusoidal pulses was chosen as an input signal. In the time range $t < 0.5$ nsec, the signal curves correspond to those of quasi-static operation (large "low" sequence). Corresponding to Figure 34, the limitation is almost reached in the range with $V_{INpp} = 0.2$ V. With fast signal changes, $t > 0.5$ nsec, the limitation is only reached above $V_{INpp} = 0.4$ V due to the narrow bandwidth of the differential amplifier. In addition it is evident that the delay time between the input and output signals decreases with increasing input voltage so that considerable jitter arises in dependence on the peak-to-peak voltage of the input signal.[33-34] This adverse effect can, as will be shown below, be

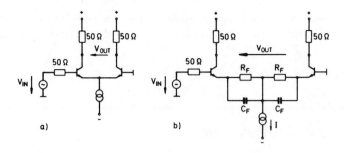

**FIGURE 35.**    (a) Differential amplifier; (b) differential amplifier with feedback.

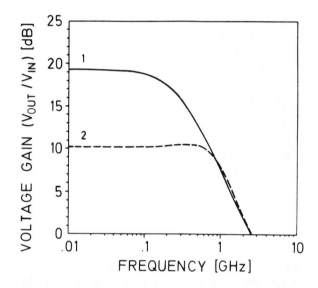

**FIGURE 36.**    Frequency response of a differential amplifier. (curve 1) Without feedback, $f_{bw}$ = 250 MHz; (curve 2) with feedback, $f_{bw}$ = 1.1 GHz.

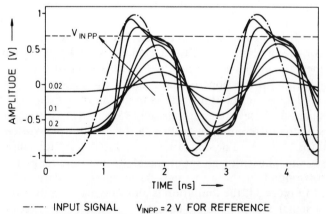

**FIGURE 37.**    Differential output signal of the differential amplifier in Figure 35a for input signals with differing peak-to-peak voltages.

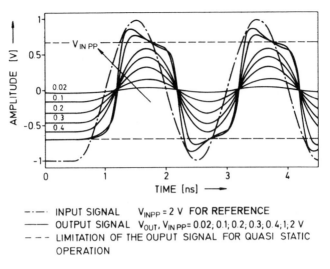

--·— INPUT SIGNAL    $V_{INPP}$ = 2 V  FOR REFERENCE
——— OUTPUT SIGNAL  $V_{OUT}$, $V_{IN PP}$= 0.02; 0.1; 0.2; 0.3; 0.4; 1; 2 V
– – – LIMITATION OF THE OUPUT SIGNAL FOR QUASI STATIC
OPERATION

FIGURE 38.    Differential output signal of the differential amplifier
in Figure 35b for input signals with differing peak-to-peak voltages.

overcome by an increase in the bandwidth: by insertion of negative feedback networks (Figure 35b) in the emitter branches, the −3-dB cut-off frequency can be raised considerably from 250 MHz to 1.1 GHz ($R_F$ = 10 Ω $C_F$ = 20 pF — Figure 36, curve 2). Of course this larger bandwidth has to be paid for by reduced gain for f → 0 and thus an increased threshold range has to be tolerated. As a result of the negative feedback network, the threshold range is increased in quasi-static operation by an amount 2 · $R_F$ · I = 300 mV (I = 15 mA, $R_F$ = 10 Ω) from 200 to 500 mV. In contrast to the differential amplifier without negative feedback, this value increases (but only by an insignificant amount) for fast signal changes as will be demonstrated with the aid for Figure 38. Corresponding to Figure 37, the signal curves in the time range t < 0.5 nsec represent quasi-static operation. It will be evident that the limitation is reached in this range as well as — in contrast to the differential amplifier without negative feedback — for fast signal changes (t > 0.5 nsec) for $V_{INpp}$ = 0.5 V. In addition it is evident that the delay time is independent of the peak-to-peak voltage of the input signal. Due to the enlarging of the bandwidth of the differential amplifier with a corner frequency far above Nyquist frequency, a constant threshold range and a delay time independent of the input signal are thus achieved.

Enlargement of the bandwidth can be achieved on the one hand by use of transistors with a higher transit frequency and on the other hand by negative feedback. On choosing negative feedback however, reduced gain and an increased but constant threshold range have to be dealt with.

The internal impedance of the signal source — $R_i$ = 50 Ω in the previous examples — exerts a relatively powerful influence on the bandwidth of the differential amplifier.

By reducing the source impedance, the bandwidth can be extended. This can be achieved by additional input emitter followers as shown in Figure 39.

In this way it is possible to increase the bandwidth to $f_{bw}$ = 1.5 GHz without reducing amplification. Figure 40 shows the frequency responses for cases with and without emitter followers and with and without negative feedback.

Due to the threshold range of 500 mV of the differential amplifier with negative feedback, the input signal must have a peak-to-peak voltage of at least 5 V to enable the penalty to remain below 1 dB, as described above in Figure 33. By arranging differential amplifiers to form a cascade it is possible to reduce the threshold range, and

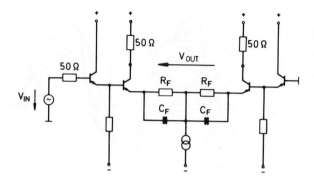

FIGURE 39.    Differential amplifier with feedback and emitter followers.

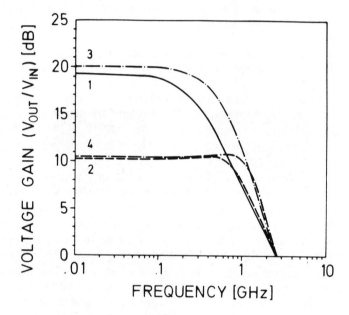

FIGURE 40.    Frequency response of a differential amplifier. (Curve 1) Without feedback, without emitter followers, $f_{bw}$ = 250 MHz; (curve 2) with feedback, without emitter followers, $f_{bw}$ = 1.1 GHz; (curve 3) without feedback, with emitter followers, $f_{bw}$ = 450 MHz; (curve 4) with feedback, with emitter followers, $f_{bw}$ = 1.5 GHz.

thus the necessary peak-to-peak input voltage too. Figure 41 shows the circuit diagram of an already implemented two-stage comparator using discrete transistors. To avoid effort (discrete mounting) no emitter followers were used, and the second stage was not set up for negative feedback. The base resistors were installed to suppress the tendency to oscillate. The threshold range of the first stage is approximately 920 mV ($2 \cdot R_F \cdot I + 200$ mV = 920 mV).

The threshold range for this entire comparator is 125 mV at 1.12 Gb/sec (560 MHz) and 200 mV at 2.24 Gb/sec (1.12 GHz). Figure 42 shows the behavior for an input signal of 560 MHz and 1.12 GHz, respectively.

### d. Emitter Degenerated Circuit as Comparator

Another possibility for implementing a comparator or limiting amplifier is provided by cascading two emitter circuits in the simplest possible arrangement (Figure 43).

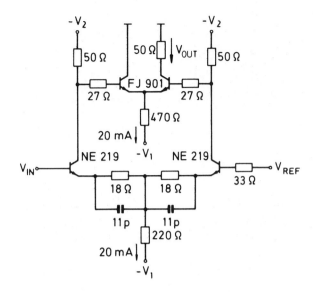

FIGURE 41. Dual-stage differential amplifier.

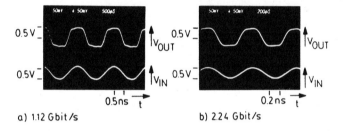

a) 1.12 Gbit/s　　　　　b) 2.24 Gbit/s

FIGURE 42. Output signal $V_{OUT}$ (upper trace) of the two-stage differential amplifier for an input signal $V_{IN}$ (lower trace) of 1.12 Gb/sec (560 MHz) and 2.24 Gb/sec (1.12 GHz).

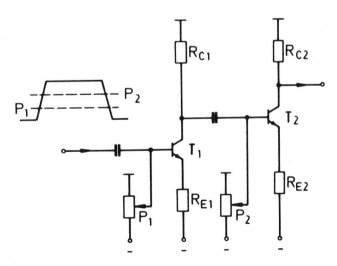

FIGURE 43. Two-stage limiting amplifier.

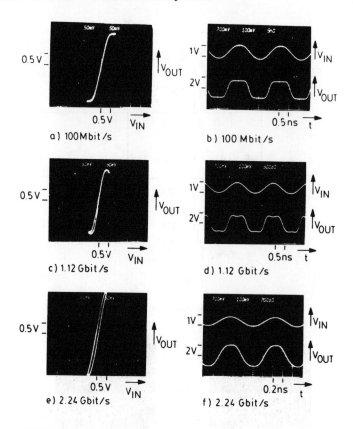

FIGURE 44.     Transfer characteristic and input and output signal of the two-stage limiting amplifier for 100 Mb/sec (50 MHz), 1.12 Gb/sec (560 MHz), and 2.24 Gb/sec (1.12 GHz).

The operating points of the transistors are adjusted in such a way that the transistor $T_1$ is nonconducting when the input signal is below the value $P_1$ and the transistor $T_2$ is nonconducting when the incoming signal exceeds the value $P_2$.

Using discrete transistors (NE 219) and a gain of 18 dB, a −3-dB cut-off frequency of $f_{bw} = 2$ GHz was obtained. The minimum threshold range obtained with two transistors is comparable with that of the single-stage differential amplifier, between 400 and 500 mV (Figure 44).

The cascading of additional stages permits a reduction in the threshold range. The oscillograms for a six-stage circuit (discrete components) for 1 Gb/sec (500 MHz) and 2 Gb/sec (1 GHz) are shown in Figure 45.

The threshold range amounts to 100 mV for 1 Gb/sec and 160 mV for 2 Gb/sec. The −3-dB cut-off frequency is > 1.5 GHz. Thus, with the bit rates under consideration here, no change occurs in the delay time dependent on the input amplitude.

Compared with the differential amplifier, this circuit concept does possess two decisive disadvantages:

1.   The circuit does not possess a direct control input in order to shift the threshold range in relation to the input signal. This can be done by means of a change in the operating points of individual transistors.
2.   In its simple configuration the circuit is not temperature-compensated. In order to reduce drift problems and influences from one stage to another, an AC-coupling should be inserted between the stages. This in turn requires a well-intermixed bit flow, in order to avoid a baseline shift within the circuit.

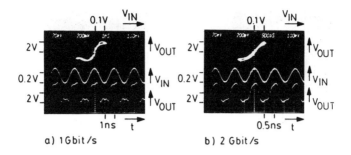

FIGURE 45.    Transfer characteristic and input and output signal for a six-stage limiting amplifier for 1 Gb/sec (500 MHz) and 2 Gb/sec (1 GHz).

### e. Integrated Comparators

At the present time integrated comparators for the Gb/sec range are not yet commercially available (1984). One of the fastest comparators available at present is the The range between $P_1$ and $P_2$ represents the threshold range. In order to obtain a high bandwidth in the threshold range, both stages have negative feedback ($R_{E1}$, $R_{E2}$). SP 9685 produced by Plessey.[35] This comparator can be used up to 300 Mb/sec. Descriptions of much faster integrated comparators are available[36-37] as well as details of the simulation of a comparator with time regeneration for 2 Gb/sec using Si bipolar technology.[38]

### 5. Timing

In the case of low-rate systems, clocked D flip-flops in integrated form are used for regenerating jitter-affected signals. As these D flip-flops are not available on the market at the present time for Gbit/sec systems, existing discrete components must be used to construct circuits consisting of fast samplers. These sample the input signal in the middle of the symbol. The RZ signal on the output side must then be converted to an NRZ signal if necessary.

We shall now consider the behavior of sampling circuits, where the signals to be sampled are no longer rectangular and where the finite duration of the sampling pulse has to be taken into account. The sampling circuit itself is described in Section II.B. Figure 46a shows a distorted, asymmetrical 2.24 Gb/sec NRZ input signal with flattened edges. Figure 46b shows the sampled RZ signal when sampled in the middle of the pulse. Signal interference of the input NRZ signal, and parts of the incompletely suppressed clock signal of the sampler lead to the lower brink of the eye diagram.

If the sampling instant is shifted, the finite width of the sampling pulse in Figure 46c and d will cause slope jitter, although the amplitudes "0" and "1" in the middle of the RZ signal are correctly reproduced. If the resulting sampled signals are to remain free of jitter, this can only be achieved by a smaller sampling pulse width — a difficult task in Gbit/sec systems. A further possibility is to shape the input signal which shall be sampled in such a way that the baseline of the "0" is free of signal interference for the duration of the sampling process: in the case of the pulses shown in Figure 46b, for about 300 psec.

Figure 47 shows the block diagram and the time curves for time regeneration of an NRZ signal affected by jitter, which is converted back into an NRZ signal following sampling.[5] The actual sampling of the inverted NRZ input signal takes place in the above-mentioned FET stage (see Figure 10). The RZ signal leaving the sampler is converted into a bipolar signal with a circuit for RZ/bipolar conversion (see Section

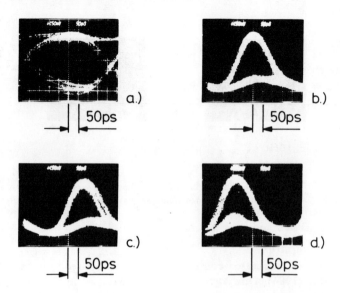

FIGURE 46.    Time regeneration of a distorted 2.24 Gb/sec NRZ signal using a sampler as shown in Figure 10a. (a) Distorted 2.24 Gb/sec NRZ input signal; (b) RZ output signal, sampling instant in center of eye; (c) sampling instant shifted to right hand brink of eye of input NRZ signal; (d) sampling instant shifted to left hand brink of eye of input NRZ signal.

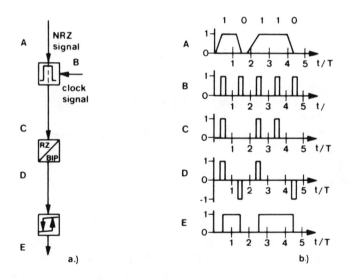

FIGURE 47.    Clock regeneration of NRZ signal with jitter. (a) Block circuit diagram; (b) diagram of time curves.

II.B.3). The Schmitt trigger at the output end is used to produce an NRZ signal (see Section II.B.4).

## 6. Complete Regeneration Circuit

Figure 48 is intended to explain the regeneration process. The center of the figure contains the block circuit diagram of the regenerator without the optoelectronic receiver (OER) or main amplifier. Eye diagrams illustrate how individual stages function; (a) shows an incoming 1.12 Gb/sec signal affected by jitter. The jitter is elimi-

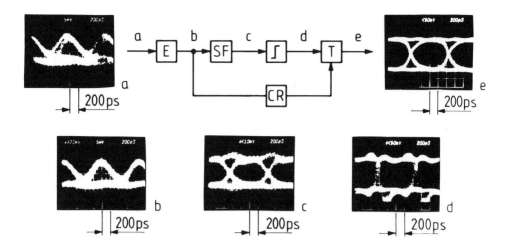

FIGURE 48.    Regeneration of 1.12 Gb/sec RZ signal. (E) Equalizer; (SF) transversal filter for signal forming; (T) time regeneration; (CR) clock regeneration.

nated with the aid of a transversal filter E (b) (Section III.B.1). The RZ signal is converted into an NRZ signal (c) using the format converter SF (Section II.B.3). The amplitude limited signal (d) (Section II.B.4) is sampled at the stage T and converted into NRZ format (e) (Section B.5). The circuits for the regeneration of 2.24-Gb/sec signals are identical with those shown in Figure 48 except the RZ/NRZ converted SF (see Chapter 10). An appropriate length of the delay line in the RZ/bipolar converter of the time regeneration circuit T must be chosen.

# REFERENCES

1. Russer, P. and Gruber, J., Hybrid-integrierter Multiplexer mit Speicherschaltdioden für den Gbit/s-Bereich, *Wiss. Ber. AEG-Telefunken,* 48, 54, 1975.
2. Barabas, U., Langmann, U., and Bosch, B. G., Diode multiplexer in the multi-Gbit/s range, *Electron. Lett.,* 14, 62, 1978.
3. Beneking, H., Filensky, W., and Ponse, F., Multi/demultiplexing in Gbit/s range using dual gate GaAs M.E.S.F.E.T.S., *Electron Lett.,* 16, 551, 1980.
4. Hagimoto, K., Ohta, N., and Nakagawa, K., 4 Gbit/s direct modulation of 1.3 μm InGaAsP/InP semiconductor lasers, *Electron. Lett.,* 18, 796, 1982.
5. Albrecht, W., Baack, C., Elze, G., Enning, B., Heydt, G., Ihlenburg, L., Walf, G., and Wenke, G., Optical digital high-speed transmission: general considerations and experimental results, *IEEE J. Quantum Electron.,* 18, 1547, 1982.
6. Hughes, J., Coughlin, B., Harbott, R., Hark, T., and Bergh, B., A versatile ECL multiplexer IC for the Gbit/s range, *IEEE J. Solid-State Circuits,* 14, 812, 1979.
7. Welbourn, P., Blau, G., and Livingstone, A., A high-speed GaAs 8 bit multiplexer using capacitor-coupled logic, *IEEE J. Solid-State Circuits,* 18, 359, 1983.
8. Rein, H.-M., Daniel, D., Derksen, R. H., Langmann, U., and Bosch, B. G., A time division multiplexer IC for bit rates up to about 2 Gbit/s, *IEEE J. Solid-State Circuits,* 19, 306, 1984.
9. Rein, H.-M. and Derksen, R., An integrated bipolar 4:1 time-division multiplexer for bit rates up to 3 Gbit/s, *Electron. Lett.,* 20, 546, 1984.
10. Baack, C., Grenzrepeaterabstand digitaler optischer Übertragungssysteme bei unterschiedlichen Laserimpulsen, *Nachrichtentech. Z.,* 30, 577, 1977.
11. Wenke, G. and Enning, B., Spectral behaviour of InGaAsP/InP 1.3 μm lasers and implications on the transmission performance of broadband Gbit/s signals, *J. Opt. Commun.,* 3, 122, 1982.
12. Beneking, H., Filensky, W., and Ponse, F., Multi/demultiplexing in Gbit/s range using dual gate GaAs MES FETs, *Electron. Lett.,* 14, 551, 1980.

13. Enning, B., Retiming 2.24 Gbit/s and demultiplexing 2.24 to 1.12 Gbit/s in an optical transmission system by means of a single-gate field effect transistor, *Electron. Lett.*, 17, 548, 1981.
14. Lindsey, W. C. and Simon, M. K., *Telecommunications Systems Engineering*, Prentice-Hall, Englewood Cliffs, N.J., 1973.
15. Byrne, C. J., Karafin, B. J., and Robinson, D. B., Systematic jitter in a chain of digital regenerators, *Bell Syst. Tech. J.*, 42, 2679, 1963.
16. Bennet, W. and Davey, J., *Data Transmission*, McGraw-Hill, New York, 1965.
17. Roza, E., Analysis of phase-locked timing extraction circuits for pulse code transmission, *IEEE Trans. Commun.*, 22, 1236, 1974.
18. Gardner, F. M., *Phaselock Techniques*, 2nd ed., John Wiley & Sons, New York, 1979.
19. Mannassewitsch, V., *Frequency Synthesizers Theory and Design*, John Wiley & Sons, New York, 1976.
20. Heydt, G., German Patent 285 4039.
21. Heydt, G. and Schiefen, G., German Patent Application, ser. 2826052.9-35, 12.6, 1978.
22. Wheeler, H. A., The interpretation of amplitude and phase distortion in terms of paired echoes, *Proc. IRE*, 27, 359, 1939.
23. Kallmann, H. E., Transversal filters, *Proc. IRE*, 28, 302, 1940.
24. Enning, B., A practical approach to the development of a transversal filter for equalization in an optical Gbit/s transmission system, *Nachrichtentech. Arch.*, 4, 333, 1982.
25. Albrecht, W. and Enning, B., A recursive filter of Gbit/s applications in an optical transmission system, *Frequenz*, 37, 82, 1983.
26. Enning, B. and Wenke, G., Influence of optical feedback on the baseband spectra of a fiber optic Gbit/s transmission system, *J. Opt. Commun.*, 4, 91, 1983.
27. Zverev, A. I., *Handbook of Filter Synthesis*, John Wiley & Sons, New York, 1967.
28. Matthaei, G. L., Young, L., and Jones, E. M. T., *Microwave Filters, Impedance-Matching Networks, and Coupling Structures*, McGraw-Hill, New York, 1964.
29. Lender, A., The duobinary technique for high speed data transmission, *IEEE Trans. Commun. Electron.*, 82, 214, 1963.
30. Enning, B., Baseline restoration in a 1.12 Gbit/s optical transmission system by means of quantized feedback equalization, *Nachrichtentech. Arch.*, 2, 183, 1980.
31. Kitami, T., Yamaguchi, H., Hohino, T., and Murata, T., An experimental 800 Mbit/s four-level amplifier compatible with the 60 MHz analog system, *IEEE Trans. Commun.*, 28, 764, 1980.
32. Waldhauer, F. D., Quantized feedback in an experimental 280 Mbit/s digital amplifier for coaxial transmission, *IEEE Trans. Commun.*, 22, 1, 1974.
33. Barna, A., Propagation delay in current-mode switching circuits, *IEEE J. Solid-State Circuits*, 10, 123, 1975.
34. Barna, A., Delay and rise time in current-mode switching circuits, *Arch. Elektron. Übertragung*, 30, 112, 1976.
35. Saul, H. P., A high speed comparator for use in optical fibre link receivers, in *7th Eur. Solid-State Circuits Conf.*, Springer-Verlag, Basel, 1981, 41.
36. Roza, E. and Millenaar, W. P., An experimental 560 Mbit/s repeater with integrated circuits, *IEEE Trans. Commun.*, 25, 995, 1974.
37. O'Connor, P., Flahive, P. G., Clemetson, W. J., Panock, R. L., Wemple, S. H., Shunk, S. C., and Takahashi, P., Monolithic multigigabit/s GaAs decision circuit for lightwave system applications, in *Conf. Opt. Fiber Commun.*, *Digest of Technical Papers*, Optical Society of America, Washington, D.C., 1984, 26.
38. Clawin, D. and Langmann, U., Monolithic multigigabit/s silicon decision circuit for applications in fibre-optic communications systems, *Electron. Lett.*, 20, 471, 1984.

Chapter 10

# REMARKS ON SYSTEM PERFORMANCE

Clemens Baack, Gerhard Elze, Bernhard Enning, and Gerhard Wenke

Numerous publications exist on the design of and investigations into high-rate optical transmission links.[1-6] The characteristics of some systems are listed in Table 1.

In System 1[9] a bit rate of 1.6 Gb/sec was transmitted over 50 km of a fiber exhibiting 0.47-dB/km attenuation. A multimode (MM) laser emitting near 1.31 $\mu$m wavelength and negligible fiber dispersion was reported for this system.

By matching the laser emission wavelength to the first-order zero-dispersion of the fiber, it was possible in System 2[7] to transmit 2 Gb/sec over 44 km of fiber with 0.57 dB/km. In the 1.5-$\mu$m range 2 Gb/sec were transmitted over 51.5 km using a zero-dispersion shifted fiber with slightly less attenuation (0.54 dB/km) (System 4[8]). Likewise in the 1.5-$\mu$m range it was possible to transmit 1.6 Gb/sec over 40 km (System 3) by means of an unshifted fiber with an attenuation of 0.35 dB/km[9] and with a longitudinal single-mode (SM) laser (DFB). With a SM C[3] laser transmission of 1 Gb/sec, over 120 km with fiber attenuation of 0.24 dB/km was achieved (System 5[10]).

It is intended here to provide information on some of the experience gathered with optical transmission links. One of the objects is to give an insight into the design of gigabit per second (Gb/sec) systems for users with limited access to optical and optoelectronic components. At the Heinrich-Hertz-Institut, Berlin, several laboratory systems, both in the optical short wavelength range at 0.85 $\mu$m[11] and also in the optical long wavelength range at 1.3 $\mu$m,[12,13] were set up and investigated. The basic structure of these systems is described in Chapter 2.

Table 2 shows the significant data of some 1.3-$\mu$m systems which have been practically implemented. They are distinguished by the bit rate transmitted, the way of the laser-fiber coupling, and the use of conventional, spectrally MM semiconductor lasers or semiconductor lasers with SM behavior achieved by an external mirror (Chapter 4). The block structure of the optical link is represented in Figure 1.

Because of the relatively high fiber attenuation of about 0.7 dB/km and transmission with lasers whose wavelengths ($\lambda$ = 1.303 to 1.31 $\mu$m for Systems 1 to 4) were not matched to the zero-dispersion of the available fiber ($\lambda_{disp.\ min.}$ = 1.375 $\mu$m), the Heinrich-Hertz-Institut laboratory systems do not provide ideal conditions regarding the achievable bit rate-distance product.

The characteristics of optoelectronic receivers (OERs) of transimpedance type (TIT) are described in Chapter 8. The calculated receiver sensitivities (bit error rate [BER] = $10^{-9}$) of a Ge avalanche photodiode (APD) with a TIT preamplifier amount to $-34.4$ dBm and $-32.0$ dBm, respectively, for the bit rates considered here: 1.12 and 2.24 Gb/sec. These bit rates correspond approximately to the highest level of the suggested European PCM (pulse code modulation) hierarchy. The measured receiver sensitivities of the unit realized using a 2-m fiber pigtail amount to $-34.3$ dBm for 1.12 Gb/sec and $-30.5$ dBm for 2.24 Gb/sec. The penalty is due to the deteriorated laser performance under modulation conditions. Comparable values are quoted.[3] Future low-noise III - V APDs suggest that even higher receiver sensitivities can be expected (Chapters 6 and 8).

Compared with the above-mentioned receiver sensitivity, a higher signal level is necessary for transmission over 21 km of a dispersive fiber for a BER = $10^{-9}$, as will be shown below. The transmission proves to be limited by interferences due to the interaction of the laser and fiber characteristics.

Table 1

**CHARACTERISTICS OF SOME 1.3 to 1.6 μm LONG-DISTANCE Gb/Sec SYSTEMS**

| System | 1 | 2 | 3 | 4 | 5 |
|---|---|---|---|---|---|
| λ [μm] | 1.31 | 1.303 | 1.54 | 1.55 | 1.55 |
| Bit rate (Gb/sec) | 1.6 | 2 | 1.6 | 2 | 1 |
| Fiber | | | | | |
|   Type | SM | SM | SM | SM(1.5 μm zero-dispersion) | SM |
|   Attenuation (db/km) | 0.47 | 0.57 | 0.35 | 0.54 | 0.27 |
|   Length (km) | 50 | 44.3 | 40 | 51.5 | 120 |
| Laser | | | | | |
|   Type | FP | BH | DFB | BH | $C^3$ |
|   Power (dBm) | −6 (pigtail) | +2.8 | −13 (pigtail) | +2.3 | +5.5 |
|   Modes | Multimode | Multimode | 1 | Multimode | 1 |
| Coupling | | | | | |
|   Structure | — | Microlens | — | Microlens | Microlens |
|   Loss (dB) | — | 4 | — | 4.5 | 4 |
| Signal | RZ | RZ | RZ | RZ | NRZ |
| Photodiode | Ge-APD $p^+nn^-$ | Ge-APD $p^+n$ | Ge-APD $p^+nn^-$ | Ge-APD $p^+nn^-$ | InGaAsP-APD |
| Bit rate × length (Gb/sec × km) | 80 | 88.6 | 64 | 103 | 120 |

Table 2

**CHARACTERISTICS OF FOUR 1.3 μm Gb/Sec SYSTEMS IMPLEMENTED AT THE HEINRICH-HERTZ-INSTITUT**

| System | 1 | 2 | 3 | 4 |
|---|---|---|---|---|
| λ [μm] | 1.31 | 1.305 | 1.303 | 1.303 |
| Bit rate (Gb/sec) | 1.12 | 2.24 | 2.24 | 2.24 |
| Fiber | | | | |
|   Type | SM | SM | SM | SM |
|   Manufacturer | SEL | SEL | SEL | SEL |
|   Attenuation (dB/km) | ∼0.7 | ∼0.7 | ∼0.7 | ∼0.7 |
|   Length (km) | 21 | 21 | 21 | 21 |
| Laser | | | | |
|   Type | BH | BH | BH | BH |
|   Power (dBm) | 0 | 0 | 2.6 | 1.9 |
|   Number of modes | 4 | 2 | 1 (external resonator) | 1 (external resonator) |
| Coupling | | | | |
|   Structure | Taper | Selfoc®/optical isolator | Taper | Selfoc®/optical isolator |
|   Loss (dB) | 3.5 | ∼6 | 3.3 | 5.8 |
| Transmitting signal | RZ | RZ | RZ | RZ |
| Photodiode | | | | |
|   Type | Ge-APD | Ge-APD | Ge-APD | Ge-APD |
|   Manufacturer | FPD 150 M Fujitsu | FPD 150 M Fujitsu | FPD 150 M Fujitsu | FPD 150 M Fujitsu |
| OER | TIT | TIT | TIT | TIT |
| Bit rate × length (Gb/sec × km) | 23.5 | 47 | 47 | 47 |
| BER = $10^{-9}$ | | | | |
|   Sensitivity limit (dBm) | (NRZ: BER ≠ $10^{-9}$) −22 | −23.7 | (NRZ: −23.4) −23.9 | −26.3 |
|   Margin (dB) (optical) | 2.9 | 2.2 | 7.6 | 6.8 |

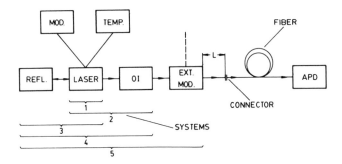

FIGURE 1. Basic structure of the optical transmission systems investigated. Numbers refer to the systems of Table 2.

On the basis of four systems (Table 2) in the 1.3-$\mu$m wavelength range, the following discussions describe the possible interference encountered in long-haul, high-rate systems and provide suggestions for overcoming such interference.

The semiconductor laser was temperature-stabilized for all systems. A system for regulating the laser output power is necessary in practical systems (on account of aging) but was not provided in the laboratory models considered here. The transmitter is directly modulated by the injection current with RZ and NRZ signals in the Gb/sec range.

Transmission of 1.12 Gb/sec over 21 km (System 1) provided a system margin of approximately 3 dB for a BER of $10^{-9}$. In this system the fiber is coupled directly to the laser by a fiber taper (see Chapter 5). With this taper high-efficiency coupling is possible. The reduced front-face reflections (short external resonator) imply less changes and interference of the laser spectrum on coupling. The coupling set-up is less sensitive to distance variations with respect to the laser spectrum and the laser threshold (see Chapter 5).[15] However, the tolerances with respect to the coupled power should be observed (e.g., channeled substrate planar [CSP] laser: side shift $< 0.7$ $\mu$m for $< 1$ dB coupling efficiency deterioration [Chapter 5]).[15]

Fiber- and connector-induced backscattering effects are not suppressed. These lead to noise side-peaks, line broadening, and excessive instability of the individual laser mode.[16] If this type of noise arises, an optical isolator must be used. It should be noted that with one Faraday isolator[17] an isolation of about 25 dB is achieved. However some index-guiding lasers (IGLs) still react to reflections of $10^{-6}$ to $10^{-8}$.

Alternatively, in the case of connector reflections, choice of a suitable pigtail length and RZ modulation can ensure that reflection of a transmitted "1" coincides with the pulsegap of the transmission signal, resulting in less distortions.

From the level diagram, Figure 2, it is obvious that the receiver sensitivity of $-34.3$ dBm at 1.12 Gb/sec is far from reached. The limiting sensitivity after 21-km transmission amounts to about $-22$ dBm. Measurements have shown that 3 to 4 dB of this deterioration can be traced back to noise arising from optical feedback.

The larger component of the sensitivity deterioration is due to laser mode partition noise (LMPN)[18-22] and bandwidth limitation of the system. The laser employed emits approximately four dominant longitudinal main modes. For System 1, the mean frequency lies approximately 65 nm outside of the fiber zero-dispersion. The correlation of the laser mode noise is abolished by propagation delay due to the material dispersion of the fiber (see Chapters 3 and 4). The achievable bit rate-distance product is restricted mainly by mode partition noise due to the width of the laser spectrum. Figure 3 shows the curve for systems in the optical long wavelength range.

Interference due to pulse overlapping (intersymbol interference) is less restricting but

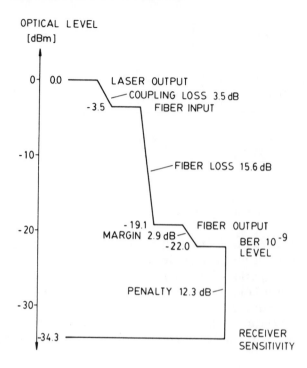

FIGURE 2.    Optical level diagram of experimental System 1 (Table 2).

likewise occurs at larger differences between the emission wavelength and the fiber zero-dispersion.[3,23] With mode jumps arising due to external influences (laser current and temperature) amounting to $\Delta\lambda$    5 nm, these burst errors limit the possible bit rate-distance product (Figure 4). The jitter and the pulse distortion of the signal resulting from mode partition noise and bursts have an effect particularly on transmission of signals in the Gb/sec range on account of their relatively small horizontal eye opening. During the decision process in the repeater, signal jitter and distortion lead to a rise in the BER. The curve of this BER in relation to the received power deviates from the typical straight line (attenuation-limited system) and exhibits a characteristic "floor", i.e., the error rate cannot be improved in this case by higher signal power.

Basically there are two possible methods of suppressing mode partition noise. A dynamic single-mode (DSM) laser can be employed (see Chapter 4) or the zero-dispersion wavelength of the fiber can be shifted to the required transmission range (e.g., 1.5 $\mu$m). This shift is made by compensation of the fiber material dispersion by the waveguide dispersion of the fiber (see Chapter 3). At present this is only possible with simultaneous deterioration of the fiber attenuation. So it is questionable whether such a shift to the 1.5-$\mu$m range is worthwhile in the long term[24,25] (Chapter 3).

In addition, fibers possessing low chromatic dispersion ( $\leq$ 2 psec/(km·nm) over the whole long-wave range between 1.3 and 1.6 $\mu$m have been implemented.[26] For example, the transmission referred to as System 2[7] (2 Gb/sec over 44.3 km) was possible although a residual material dispersion of 2.2 psec/km·nm was estimated. Wavelength multiplex operation is also possible in this range with conventional broad-spectrum lasers. Mention should be made of the fact that partition noise is also generated by spectral filtering of the laser modes, e.g., in wavelength division multiplex (WDM) systems at the slope of the demultiplexer transmission band and by filtering of the polarization state.[27,28]

As shown in Chapter 4, intrinsic SM behavior of the laser (approximately 90% of

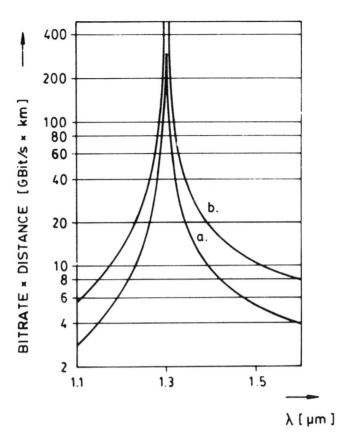

FIGURE 3. Bit rate-distance product caused by mode partition noise (worst case K = 1)[20] against wavelength. RMS half-width of source is a = 2 nm, b = 1 nm. The wavelength of minimum dispersion is assumed to be 1.3 μm.[23]

the output power is concentrated in one mode) can be maintained in the Gb/sec range by RZ modulation and biasing the laser slightly above threshold.[29] The spectral behavior is by this means improved — to an extent, however, at the expense of the extinction ratio.

In System 2[13] it was possible to transmit a bit rate of 2.24Gb/sec with a BER of $10^{-9}$ by use of an optical isolator with 25-dB isolation and about 6-dB insertion loss and by weak mode selection (Chapter 4) by means of an adjustable Selfoc® coupling lens in front of the laser facet.

In System 3, a laser mode was selected by use of a short external resonator at the laser back facet[30-32] (Chapter 4). The neighboring modes are suppressed by about 20 to 25 dB. If success is achieved in setting up such a multicavity structure as a long-term stable system, possibly by inclusion of a regulation system for the reflector distance[33] or the laser parameter current or temperature, a technically simple arrangement for laser mode selection will be available. A monolithic component, e.g., a distributed feedback/distributed Bragg reflector (DFB/DBR) laser (see Chapter 4) is however, preferable to the above arrangement, assuming that the neighboring mode suppression is likewise approximately equal to 25 dB.

The level diagram of System 3 is shown in Figure 5. The received eye patterns of 1.12 and 2.24 Gb/sec signals are shown in Figure 6 after 21-km transmission, using a wavelength approximately 72 nm outside of the fiber zero-dispersion. The much less

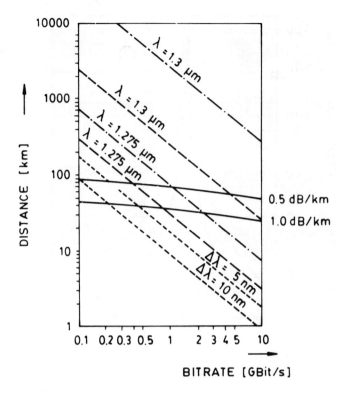

FIGURE 4.    Theoretical limits of repeater span against bit rate for various kinds of system limitation. RMS half-width of laser is 2 nm (fiber minimum dispersion at 1.3 μm). Solid lines indicate limitation due to fiber attenuation. Dot-dash lines indicate fiber bandwidth limitation. Dashed lines indicate mode partition limitation (K = 1).[2] Dotted lines indicate limitation by burst errors due to mode jumping.[23]

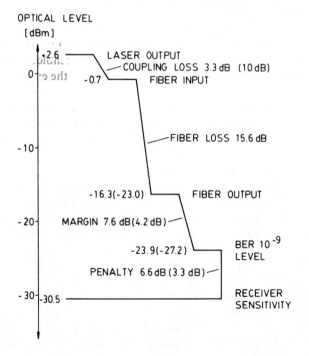

FIGURE 5.    Optical level diagram of experimental System 3.

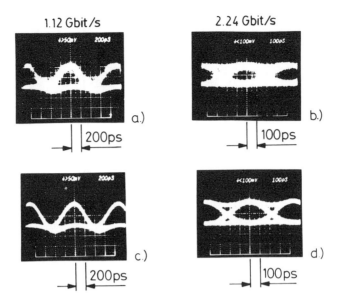

FIGURE 6. Eye patterns received after 21-km SM fiber transmission for 1.12 and 2.24 Gb/sec RZ signals. (top) Conventional 1.3-$\mu$m buried heterostructure (BH) laser with RMS half-width $\Delta\lambda \cong 2$ nm; (below) 1.3-$\mu$ BH laser with external cavity and side mode suppression of 25 dB. Due to laser bandwidth limitation, NRZ signals are received in the case of 2.24 Gb/sec.

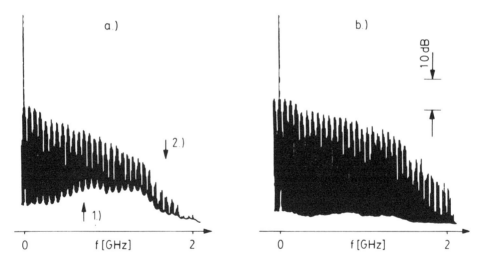

FIGURE 7. Power density spectrum received in the baseband of a 2.24 Gb/sec signal (see text). (a) 1. Periodic noise enhancement (inverse round-trip time); 2. bandwidth reduction (the physical causes have not been completely explained). On use of a conventional optical connector with Fresnel reflection. (b) Undistorted signal on use of an optical connector without airgap (physical contact type).

critical eye patterns for transmission with a DSM laser (short external reflector) permit error-free transmission (BER < $10^{-9}$).

Such an arrangement will not suppress interference due to fiber backscatter or connector reflections.[28] The periodic noise enhancement in the baseband spectrum resulting from connector reflections as seen in Figure 7a and 7b shows that on use of a connector without an airgap (i.e., with physical contact of the fiber cores) this interference no longer occurs.

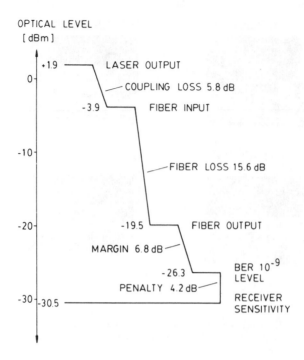

OPTICAL LEVEL
[dBm]

+1.9 ——— LASER OUTPUT

0 —

‚— COUPLING LOSS 5.8 dB

-3.9 ⌐———  FIBER INPUT

-10 —

‚—FIBER LOSS 15.6 dB

-20 —    -19.5 ⌐———  FIBER OUTPUT

MARGIN 6.8 dB ⌐

BER 10⁻⁹
LEVEL

-26.3

PENALTY 4.2 dB ⌐

-30 —  -30.5 ————————  RECEIVER
                        SENSITIVITY

FIGURE 8.    Optical level diagram of experimental System 4
(Table 2).

Analysis of the system behavior under the influence of interference by means of the power density spectrum in the baseband has proved to be very helpful in practice. Measurements should be taken under real modulation conditions with a pseudo-statistic signal.[28] Figure 7 shows the measured power density spectra of a 2.24 Gb/sec RZ signal. Corresponding to the repetition rate of the pseudo-statistic pattern and the processing of the modulating signal,[4] a discrete line spectrum is produced with a line spacing of 8.82 MHz and additional zeros at a spacing of 62.22 MHz. If the bandwidth of the spectrum analyzer is suitably selected, the signal and noise components of the signal can be observed simultaneously. The envelope on the upper side characterizes the signal spectrum. Between the discrete lines there are no signal components, so the lower envelope shows the noise components of the signal. This method permits noise sources to be detected and overcome. Quantification is achieved by measurement of the error rate.

In System 4 (level diagram in Figure 8), mode selection was achieved using an external reflector and an optical isolator of approximately 25-dB isolation. Compared with System 3, i.e., mode selection alone, a sensitivity decreased by about 2.5 dB was achieved by suppression of the feedback interference. However, this did not produce a higher system margin since the larger insertion loss of the optical isolator (5.8 dB) compared with 3.3 dB in the case of taper coupling has to be taken into account. A system penalty (difference between the sensitivity achieved by transmission over the fiber and the receiver sensitivity) of approximately 4 dB remains. This is due to inadequate side mode suppression (37 dB has been achieved),[10] residual noise due to the fiber coupling process, and a nonideal extinction ratio (compromise between narrow spectrum and the underlaying constant light level). In practical systems, consideration must always be given to whether the influence of partition noise (mode selection, zero-dispersion matching, use of components without λ-dependent transmission) or the influence of optical feedback interference (optical isolation, weak laser-to-fiber cou-

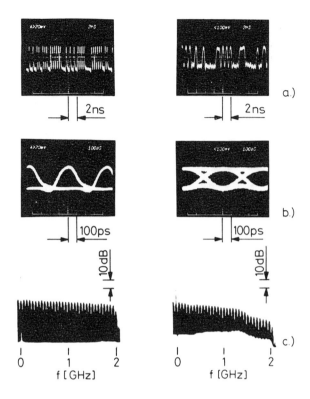

FIGURE 9.  2.24 Gb/sec RZ transmission over 21 km (System 3). (left) Driving signal; (right) received signal (a) signal (b) eye pattern (c) baseband spectrum.

pling, suitable pigtail length) predominates. The counter-measures to be adopted should be chosen accordingly.

Figure 9 and 10 show the transmission signals and the received signals of System 3 for RZ and NRZ transmission. Both types of modulation are suitable (approximately same sensitivity is achieved) for transmission with the mode-stabilized transmission unit. The time signals, the eye pattern diagrams, and the baseband spectra are shown.

With direct modulation of the laser, a slight deterioration in the laser mode selection occurs[32] (Chapter 4). Likewise, direct modulation leads to line widening and noise.[34] Particularly at 1.5 μm, bandwidth restrictions occur due to line widening.[35] As is well known, such interference can be avoided by external modulation.[36,37]

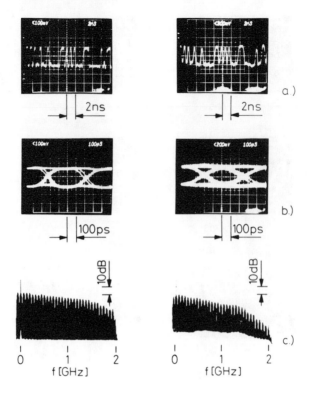

FIGURE 10.    2.24 Gb/sec NRZ transmission over 21 km (System 3). (left) Driving signal; (right) received signal (a) signal (b) eye pattern (c) baseband spectrum

# REFERENCES

1. Yamada, J. I., Saruwatari, M., Asatani, K., Tsuchiya, H., Kawana, A., Sugiyama, K., and Kimura, T., High-speed optical pulse transmission at 1.29 μm wavelength using low-loss single-mode fibers, *IEEE J. Quantum Electron.*, 14, 791, 1978.

2. Yamada, J. I. and Kimura, T., Characteristics of Gbit/s optical receiver sensitivity and long-span single-mode fiber transmission at 1.3 μm, *IEEE J. Quantum Electron.*, 18, 718, 1982.

3. Yamada, J. I., Kawana, A., Miya, T., Nagai, H., and Kimura, T., Gigabit/s optical receiver sensitivity and zero-dispersion single-mode fiber transmission at 1.55 μm, *IEEE J. Quantum Electron.*, 18, 1537, 1982.

4. Albrecht, W., Baack, C., Elze, G., Enning, B., Heydt, G., Ihlenburg, L., Walf, G., and Wenke, G., Optical digital high-speed transmission: general considerations and experimental results, *IEEE J. Quantum Electron.*, 18, 1547, 1982.

5. Baack, C., Elze, G., Grosskopf, G., and Walf, G., Digital and analog optical broad-band transmission, *Proc. IEEE*, 71, 198, 1983.

6. Tsang, W. T., Logan, R. A., Olsson, N. A., Temkin, H., van der Ziel, J. P., Kaminow, I. P., Kasper, B. L., Linke, R. A., Mazurczyk, V. J., Miller, B. I., and Wagner, R. E., 119 km, 420 Mb/s Transmission with a 1.55 μm single-frequency laser, Paper 9, in Proc. OFC, New Orleans, 1983.

7. Yamada, J. I., Machida, S., and Kimura, T., 2 Gbit/s Optical transmission experiments at 1.3 μm with 44 km single-mode fibre, *Electron. Lett.*, 17, 479, 1981.

8. Yamada, J., Kawana, A., Nagai, H., and Kimura, T., 1.55 μm Optical transmission experiments at 2 Gbit/s using 51.5 km dispersion-free fibre, *Electron. Lett.*, 18, 98, 1982.

9. Nakagawa, K., Otha, N., and Hagimoto, K., Design and performance of an experimental 1.6 Gbit/s optical repeater, Paper 27C2-1, in Proc. IOOC, Tokyo, 1983.

10. Linke, R. A., Kasper, B. L., Campbell, J. C., Dentai, A. G., and Kaminow, I. P., A 1 Gbit/s lightwave transmission experiment over 120 km using a heterojunction APD receiver, Post deadline paper WJ7, in Proc. OFC, New Orleans, 1983.

11. Albrecht, W., Baack, C., Elze, G., Enning, B., Heydt, G., Peters, K., Walf, G., and Wenke, G., 2.24 Gbit/s Optical transmission system at 0.85 μm wavelength, *Electron. Lett.*, 17, 664, 1981.

12. Baack, C., Elze, G., Enning, B., Walf, G., and Wenke, G., Optical transmission of 16 digitalized TV channels over 21 km fibre at a bit rate of 1.12 Gbit/s, *Electron. Lett.*, 17, 517, 1981.

13. Albrecht, W., Elze, G., Enning, B., Walf, G., and Wenke, G., Experiences with an optical long-haul 2.24 Gbit/s transmission system at a wavelength of 1.3 μm, *Electron. Lett.*, 18, 746, 1982.

14. Linke, R. A., Kasper, B. L., Jopson, J., Campbell, C., Dentai, A. G., Tsang, W. T., Olsson, N. A., Logan, R. A., Johnson, L. S., and Henry, C. H., A 2 Gbit/s 71 km transmission experiment using a 1.5 μm ridge-type distributed feedback (DFB) laser, in 9th IEEE Int. Semicond. Laser Conf., Rio, Session E-5, 1984.

15. Wenke, G. and Zhu, Y., Comparison of efficiency and feedback characteristics of techniques for coupling semiconductor lasers to single-mode fiber, *Appl. Opt.*, 22, 3837, 1983.

16. Miles, R. O., Dandridge, A., Tveten, A. B., Taylor, H. F., and Giallorenzi, T. G., Feedback-induced line broadening in CW channel-substrate planar laser diodes, *Appl. Phys. Lett.*, 37, 990, 1980.

17. NEC 1.3 μm Band Optical Isolator OD-8313, Data Sheet, 1983.

18. Okano, Y., Nakagawa, K., and Ito, T., Laser mode partition noise evaluation for optical fiber transmission, *IEEE Trans. Commun.*, 28, 238, 1980.

19. Shimada, S., Systems engineering for long-haul optical fiber transmission, *Proc. IEEE*, 68, 1304, 1980.

20. Ogawa, K., Analysis of mode partition noise in laser transmission systems, *IEEE J. Quantum Electron.*, 18, 849, 1982.

21. Grau, G., Schwankungserscheinungen als prinzipielle und praktische Leistungsgrenzen optischer Nachrichtensysteme, *AEÜ*, 37, 137, 1983.

22. Grosskopf, G., Küller, L., and Patzak, E., Laser mode partition noise in optical wideband transmission links, *Electron. Lett.*, 18, 493, 1982.

23. Ogawa, K., Considerations for single-mode fiber systems, *Bell Syst. Tech. J.*, 61, 1919, 1982.

24. Midwinter, J. E., Monomode fibres for long-haul transmission systems, *Br. Telecom Technol. J.*, 1, 5, 1983.

25. Stanley, J. W., Hooper, R. C., and Smith, D. W., The performance of experimental monomode transmission links, in Proc. 8th ECOC, Cannes, A XIV, 1982.

26. Cohen, L. G., Mammel, W. L., and Jang, S. J., Low-loss quadruple-clad single-mode lightguides with dispersion below 2 ps/(km·nm) over the 1.28 — 1.65 μm wavelength range, *Electron. Lett.*, 18, 1023, 1982.

27. Tomita, A., Duff, D. G., and Cheung, N. K., Mode partition noise caused by wavelength-dependent attenuation in lightwave systems, *Electron. Lett.*, 19, 1079, 1983.

28. Enning, B. and Wenke, G., Demonstration of coloured noise and signal distortion in baseband spectra of broadband optical transmission systems, *NTZ-Arch.*, 5, 301, 1983.

29. Wenke, G. and Enning, B., Spectral behavior of InGaAsP/InP 1.3 μm lasers and implications on the transmission performance of broadband Gbit/s signals, *J. Opt. Commun.*, 3, 122, 1982.

30. Preston, K. R., Woollard, K. C., and Cameron, K. H., External cavity controlled single longitudinal mode laser transmitter module, *Electron. Lett.*, 17, 931, 1981.

31. Cameron, K. H., Chidgey, P. J., and Preston, K. R., 102 km Optical fibre transmission experiments at 1.52 μm using an external cavity controlled laser transmitter module, *Electron. Lett.*, 18, 650, 1982.

32. Elze, G., Grosskopf, G., Küller, L., and Wenke, G., Experiments on modulation properties and optical feedback characteristics of laser diodes stabilized by an external cavity or injection locking, *IEEE J. Lightwave Technol.*, LT-2, 1063, 1984.

33. Preston, K. R., Simple spectral control technique for external cavity laser transmitters, *Electron. Lett.*, 18, 1092, 1982.

34. Kishino, K., Aoki, S., and Suematsu, Y., Wavelength variation of 1.6 μm wavelength buried heterostructure GaInAsP/InP lasers due to direct modulation, *IEEE J. Quantum Electron.*, 18, 343, 1982.

35. Olsson, N. A., Dutta, N. K., and Liou, K. Y., Dynamic linewidth of amplitude-modulated single longitudinal mode semiconductor lasers operating at 1.5 μm wavelength, *Electron. Lett.*, 20, 121, 1984.

36. Leonberger, F. J., Progress in Ti:LiNbO₃ and InP waveguide devices for signal processing applications, Paper 30B2-1, in Proc. IOOC, Tokyo, 1983.

37. Alferness, R. C., Korotky, S. K., Jayner, C. H., and Buhl, L. L., 8 Gbit/s Optical signal encoding at 1.32 μm with a Ti:LiNbO₃ waveguide directional coupler modulator, Paper 30B2-2, in Proc. IOOC, Tokyo, 1983.

Chapter 11

# FUTURE ASPECTS

Clemens Baack, Gerhard Elze, Günter Heydt, and Gerhard Wenke

## TABLE OF CONTENTS

I.    Introduction.................................................................................196

II.   Development Towards Higher Bit Rates .............................................196

III.  Integrated Optics..........................................................................196

IV.   Optical Heterodyne Detection .........................................................197

V.    Very Long Optical Wavelength Range...............................................199

VI.   Possible Applications ....................................................................199
      A.    Single-Mode Systems for Subscriber Lines of Public Networks .......199
      B.    Subscriber Lines with High Bit Rates .......................................200
      C.    Integrated Optics in the Subscriber Field...................................201
      D.    Optical Heterodyne Detection in the Subscriber Sphere.................201
      E.    Coherent Optical Communications for Long-Haul Systems in
            Public Networks ................................................................202

References ............................................................................................203

# I. INTRODUCTION

This chapter discusses development trends already predictable for broadband (BB) optical data transmission and their possible repercussions on future communications systems. Such trends primarily involve the further development of existing direct-detection systems to achieve higher bit rates, the use of optical heterodyne methods, and the introduction of monolithic integrated optoelectronic components. Implementation of the results of these developments is conceivable in future public communications systems, not only for long-haul systems, but also within the subscriber sphere. It should however be stressed that the essential condition for implementation of these technologies involves the employment of single-mode (SM) fibers.

# II. DEVELOPMENT TOWARDS HIGHER BIT RATES

At present, experimental laboratory systems with bit rates up to the 2-Gb/sec range already exist in several research laboratories. At the same time work is being carried out on the development of components which will be suitable for even higher bit rates. Preliminary experiments have shown that direct modulation up to 10 Gb/sec is possible with special laser structures.[1] Photodiodes for bandwidths of up to 10 GHz are already commercially available. Very high-speed integrated electronic circuits are likewise being developed including, for instance, Si circuits for fields of application up to 3 Gb/sec[2] and GaAs circuits with an operating speed of over 4 Gb/sec.[3]

For optical systems with such high bit rates, the demands discussed in Chapters 4 and 10 to be placed on the spectrum of the light signal coupled into the fiber from the transmission module apply to an even greater extent. Dynamic single mode (DSM) lasers are thus prerequisite together with coupling methods which suppress optical feedback.

# III. INTEGRATED OPTICS

The goal of integrated optics is to integrate both active and passive optical and electronic components together on a substrate so that a monolithic whole is produced (optoelectronic integrated circuits, OEICs). Active optical components such as lasers and photodiodes serve for electronic-to-optical and optical-to-electronic conversion; passive optical components such as waveguides, gratings, lenses, filters and couplers serve for guiding, refracting, imaging, and spectral splitting as well as switching light; electronic components such as transistors and resistors serve to control electronic devices and for electronic signal processing.

Integrated optics has certain parallels to microelectronics which involves implementation of highly complex electronic systems on small, sturdy, and reliable Si chips which can be produced economically in large quantities. Accordingly, integrated optics is directed towards developing optoelectronic systems as small, sturdy, and reliable semiconductor chips which can likewise be produced economically in large quantities. However, because of the larger dimensions of basic optical components, far fewer components can be accommodated on one chip in integrated optics than in microelectronics. Since only component assemblies of low complexity are required for optical BB transmission technology, this aspect is of no significance for the present considerations.

The indirect band structure of Si prohibits the development of light sources such as lasers on a Si basis. Therefore, for the development of OEICs, heterojunction semiconductors of the III to V chemical element material system are applied. OEICs are currently developed on a GaAs basis for the optical short-wave range and on an InP basis for the optical long-wave range.

Apart from work on the development of integrated optics on a semi-conductor basis, considerable effort is being expended all over the world on the development of so-called "hybrid optics". In this field lithium niobate or related materials are employed, which permit production of low-loss, passive, optical components, exhibit relatively large electrical, magnetic, and acousto-optical effects, and are technologically easier to handle and cheaper to produce than semiconductors. Lasers, photodiodes, and electronic components cannot however, be produced with these materials, so that such components have to produced separately and connected to the substrate at a later stage. Hybrid optics is thus comparable with electronic thick and thin film circuit technologies in which discrete components such as transistors and diodes are subsequently incorporated in the circuit. Similar to electronic thick and thin film circuits, hybrid optical circuits are cost-effective up to medium-size production runs. On the other hand, as is the case with electronics, monolithic integration is essential for large quantities.

Hybrid optics as a technology is already accessible today. It will play an important role in data processing technology in coming years. On the other hand, it will take several years of development in the field of III to V semiconductor materials before cost-effective mass production of OEICs is possible. The significance of integrated optics in public and local networks, in sensor technology, and in high-speed signal processing is described.[4]

## IV. OPTICAL HETERODYNE DETECTION

In optical communications, work is at present being carried out with the aim of applying the principle of heterodyne reception, which has proved its advantages over decades in radio technology, to optical detection. This promises the following features:

- Increase of 10 to 20 dB in receiver sensitivity
- High receiver selectivity
- Possibilities of selecting a channel by tuning the local laser

Figure 1 shows the principle of an optical heterodyne transmitter: the received light signal and the light of the local laser with wave magnitudes $\underline{a}_s$ and $\underline{a}_L$ are fed via a directional coupler (coupling factor k) to an optical detector with an output current proportional to the light power (e.g., PIN diode):

$$\underline{a}_S = \sqrt{P_S}\, e^{j(\omega_S t + \Phi_S)} \qquad \underline{a}_L = \sqrt{P_L}\, e^{j(\omega_L t + \Phi_L)} \tag{1}$$

where $P_s$, $\omega_s$, $\Phi_s$ and $P_L$, $\omega_L$, $\Phi_L$ are the powers, angular frequencies, and instantaneous phases of the signal and local laser waves. Neglecting the coupler losses, the detector current at the output of a PIN diode is as follows:

$$i_{het} = \frac{e\eta}{hf} \{(1 - k^2)\, P_S \\ + k^2 P_L + 2k\sqrt{(1 - k^2)P_S P_L}\, \cos[(\omega_S - \omega_L)t + (\Phi_S - \Phi_L)]\} \tag{2}$$

The first two terms provide a signal in the baseband; the third term provides a signal at an intermediate frequency $\omega_s - \omega_L$. Conversion to the intermediate frequency range is proportional to the product of the wave magnitudes as long as provisions are provided to ensure that $\underline{a}_s$ and $\underline{a}_L$ have the same field shape and polarization, i.e., that they

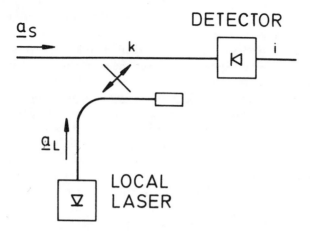

FIGURE 1.    Principle of optical heterodyne detection.

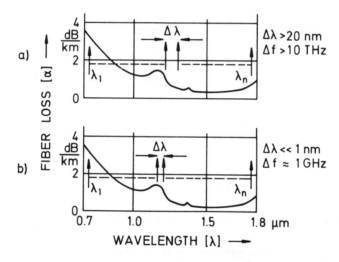

FIGURE 2.    Channel allocation in a SM fiber. (a) WDM; (b) optical frequency multiplex with heterodyne detection.

are spatially coherent. It is likewise necessary to stabilize the laser frequencies so that a constant intermediate frequency is achieved. If the phase is also to be used as an information carrier (coherent optical transmission), the phase difference $\Phi_s - \Phi_L$ will have to be sufficiently constant for the measuring time (e.g., for the time taken to transmit 1 b). For $k = 0$, Equation 2 gives the familiar relationship for the detector current $i_s$ with baseband direct reception:

$$i_d = \frac{e\eta}{hf} P_s$$

Comparison with the intermediate frequency term in Equation 2 will show that the sensitivity in relationship to direct reception rises with the light power of the local laser. The limit of this gain, depending on the wavelength of the light, amounts to 10 to 20 dB and is due to the noise of the local laser.

In optical communications technology, the capacity of transmission systems can be increased by application of λ-multiplex technology at which several light carriers of different wavelengths are transmitted via the fiber and are separated at the output by optical filters (Figure 2). To ensure reliable separation, the light carriers must be at least several tens of nanometers apart with the present state of development of filter and laser technology. Thus a maximum of about 40 light carriers can be accommodated with a wavelength difference of, e.g., 30 nm (corresponding to approximately 10 THz), in the low-loss wavelength range of a modern SM fiber of 0.7 to 1.8 $\mu$m (Figure 2b). Optical $\lambda$-multiplex technology corresponds to direct detection in radio broadcasting technology in as far as in both cases the signal received is filtered and detected without prior frequency conversion. In the case of optical heterodyne detection, the intermediate frequency can be located in the microwave range in which in comparison to the frequency of the light carrier, very narrow band filters can be implemented (Figure 2c). Thus, carrier spacings of, e.g., 1 GHz can be achieved. From a purely arithmetic point of view, the transmission capacity of one SM fiber is about 250,000 channels. It is thus not limited by the bandwidth but by nonlinear effects of the transmission medium.

## V. VERY LONG OPTICAL WAVELENGTH RANGE

As described in Chapter 3, optical fibers constructed of $SiO_2$ have minimum losses at about 1.55 $\mu$m. The loss due to infrared absorption of the material increases considerably towards longer wavelengths. Since the Raleigh dispersion falls with $\lambda^{-4}$, lower-loss waveguides are to be expected at longer wavelengths as a basic principle. Thus for several years attempts have been made in various laboratories to produce light waveguides of materials whose absorption edges lie at longer wavelengths than with $SiO_2$. The materials used can be divided into two categories: glass and crystal. In recent years steady advances have been made in reducing losses with waveguides of both materials. Typical values now lie between 5 and 100 dB/km in the wavelength range from 2 to 10 $\mu$m and are thus still far removed from the losses predicted which lie at $10^{-2}$ to $10^{-3}$ dB/km. Suitable laser and photodiodes for this wavelength range are still not available for system applications. On account of the low energy gap of the semiconductor materials which are in principle suitable for this purpose, the development of components which are suitable for operation at room temperature is unlikely in the future. However, there may be application fields in the far future, e.g., for repeaterless bridging of long undersea distances.

## VI. POSSIBLE APPLICATIONS

Following are some examples of application possibilities of the newly mentioned technologies for future public and private networks. It may be pointed out that basic developing efforts are carried out at present in order to investigate the chances to realize such system concepts.

### A. Single-Mode Systems for Subscriber Lines of Public Networks

High-rate, SM systems as already discussed are at present mainly conceived for use in long-haul purposes in public communications networks.

One essential condition for the use of SM fiber in the subscriber field involves the necessity to overcome existing problems involving splicing and plug connectors. Advances in splicing methods observed in recent years would suggest that splicing problems will be solvable in the future. Convincing proposals for sturdy, reliable, and cheap SM connectors with low transmission losses which can be mass produced are at present still not in sight. Solutions have, however, been published for SM connectors which operate reliably with beam broadening when low demands are placed on the precision. The higher transmission losses encountered with these are tolerable on account of the system reserves existing at the subscriber level due to short line lengths.

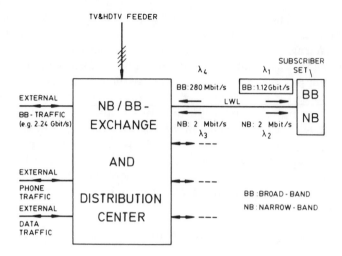

FIGURE 3. Schematic representation of an experimental BB system for the local network.

Thus in the distant future, use of SM fibers would appear practically possible. This would bring about the following advantages:

1.  Operation of subscriber connection lines at high bit rates offers high system flexibility as regards transmission for BB services.
2.  Possibilities are opened up for use of monolithic OEIC components.
3.  By the possible use of optical heterodyne detection and of OEIC components, interesting aspects arise for distributing BB services in the subscriber field.
4.  An opportunity is created for uniform fiber technology for all network levels.

Examples of the advantages mentioned in 1, 2, and 3 will be given in the following section.

## B. Subscriber Lines with High Bit Rates

The SM subscriber line shown in Figure 3 permits transmission of high bit rates (1.12 Gb/sec) from the exchange to the subscriber. The subscriber can be offered up to 15 TV programs by present-day standards with 70 Mb/sec per program or several programs in accordance with some future high-definition TV (HDTV) standard[5] with 280 Mb/sec per program.

In addition to the problem of preventing the implementation of optoelectronic converters for 1.12 Gb/sec from becoming too elaborate, the problem of high-rate electronics in the subscriber field must be treated here. The transmitter-side, speed-critical component stage, a multiplexer 280/1120 Mb/sec, was implemented by monolithic integration in this experimental system.[6] The reception-side, speed-critical component stage, a channel selector 70 from 1120 Mb/sec or 280 from 1120 Mb/sec for HDTV, was implemented as a thin film circuit (1 in. × 2 in.) which exhibits a power consumption of 1.5 W.[7] Monolithic integration is possible.

In this system, narrow band services are dealt with via separate wavelengths ($\lambda_2$, $\lambda_3$) with 2 Mb/sec. As a result of this, e.g., the emergency power supply of the telephone service is made easier.

The wavelength $\lambda_4$ serves for video telephone transmission from the subscriber to the exchange. Either four video calls by the present video standard (70 Mb/sec), or in the distant future, one video call by the HDTV standard can be dealt with.

BB communications between the exchanges will use 2.24-Gb/sec links.

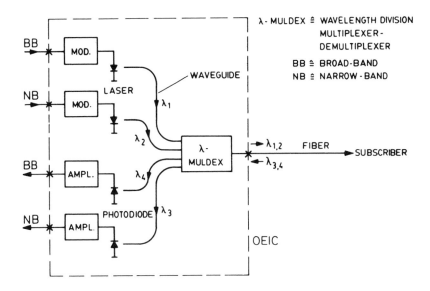

FIGURE 4.   EO/OE transducer for a subscriber station.

## C. Integrated Optics in the Subscriber Field

On introduction of optical communications technology in the public network, large quantities of optical components must be expected at the subscriber level, i.e., there is a pressing need for the use of integrated optical components.

As an example, Figure 4 shows the exchange-side EO/OE converters of the system in accordance with Figure 3. A complementary converter is employed on the subscriber side. Converters of this type are at the present time assembled from individual components. The optical multiplexer/demultiplexer ($\lambda$-muldex) is a micro- or fiber-optical component which is very demanding from the point of view of fine mechanics.

The emission wavelengths of the lasers are to be carefully tuned to the transmission ranges of the multiplexer. For this purpose, in the long term, a small, sturdy, and reliable OEIC which can be mass produced under cost-effective conditions is to be aimed at.

## D. Optical Heterodyne Detection in the Subscriber Sphere

As already mentioned, the optical heterodyne receiver is highly selective. It is possible to transmit a large number of light carriers at very close spacing (e.g., several gigahertz) through one fiber. If, for example, it is intended to offer 100 TV channels to all the subscribers in a local network, one light carrier is allocated to each channel. The carriers of all 100 channels ($\lambda_1$ to $\lambda_{100}$ in Figure 5) are transmitted via the subscriber line. Each subscriber can select the required channel with the aid of a tunable optical heterodyne receiver. For this purpose the local laser is tuned in such a way that its light together with the light of the relevant carrier produces the intermediate frequency.

It will not be possible to produce optical heterodyne receivers economically in a hybrid form. This interesting solution for an optical distribution network will only be implementable when optical heterodyne receivers are available as OEICs.

The distribution system could be operated with ASK modulation and envelope detection for which purpose the DSM lasers already described, e.g., distributed feedback (DFB) or distributed Bragg reflector (DBR) lasers with their relatively large line widths of several tens of megahertz, can be used.[8]

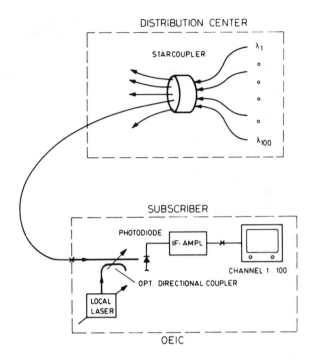

FIGURE 5.    BB distribution network with heterodyne detection.

## E. Coherent Optical Communications for Long-haul Systems in Public Networks

In long-haul systems for public networks and in particular for maritime cables, it is intended to take advantage not only of the high selectivity of optical heterodyne receivers for transmission of a large number of carriers with close spacing, but also of their high sensitivity for increasing the range. Instead of ASK modulation, which only uses the amplitude of light, FSK, PSK, or DPSK modulation techniques are chosen which make use of the phase of the light (coherent optical communications technology). For this purpose transmitter and local lasers are required which are not only of SM type but whose line width lies in the kilohertz range. Apart from gas lasers, this demand is fulfilled at present by semiconductor lasers with "long" external resonators (narrow line width [NLW] lasers).

Future BB networks will offer the subscriber a video telephone system with high picture quality. As a result, an extremely heavy flow of data on the long-haul level of these networks will arise. To deal with such heavy flows of data, it would be possible, e.g., to lay a large number of SM fibers parallel to each other in which connection each fiber is subject to multi-use applications due to employment of wavelength division multiplex (WDM) techniques and high bit rates. This leads to very elaborate repeater stations as a result of the large number of fibers, the WDM technique, and the high bit rates. Coherent optical communications techniques may well offer a way out of this difficulty. As shown in Figure 6, several light carriers are transmitted per fiber which, in contrast to the WDM technique, have spacing of only a few gigahertz so that, e.g., five carriers ($\lambda_1..._5$) can be amplified together by one optical amplifier. As a result of this, no multiplexer/demultiplexers as well as no OE and EO converters are to be incorporated in the repeater stations. The carriers are selected at the end of the long-haul link by heterodyne receivers.[4]

Initially such a system could be operated with DSM lasers with large line width, e.g., DFB lasers, and with ASK modulation. In the distant future, however, NLW lasers will be employed to achieve maximum ranges by coherent methods of modulation.

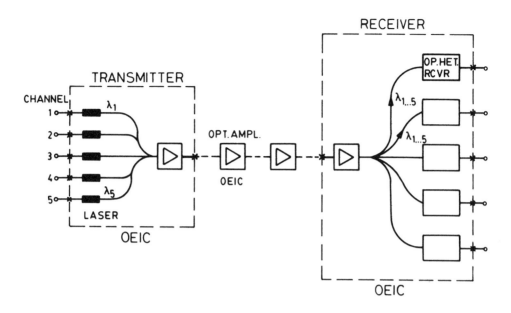

FIGURE 6. Model of a future long-haul transmission system with optical amplifiers and heterodyne detection.

The long-haul link shown in Figure 6 which does not operate with conventional present-day electronic repeaters but with optical amplifiers will only be economically implementable when the transmitter unit, the optical amplifiers, and the receiver units are available as OEICs.

## REFERENCES

1. Lau, K. Y., Bar-Chaim, N., and Ury, I., A 11 GHz direct modulation bandwidth GaAlAs window laser on semi-insulating substrate operating at room termperature, post deadline Paper W71, in Conf. Opt. Fiber Commun., Digest of Technical Papers, Optical Society of America, Washington, D.C., 1984.
2. Rein, H. M. and Derksen, R. H., Integrated bipolar 4:1 time-division multiplexer for bit rates up to 3 Gbit/s, *Electron. Lett.*, 20, 546, 1984.
3. Liechti, C. A., Baldwin, G. L., Gowen, E., Joly, R., Namjoo, M., and Podell, A. F., A GaAs MSI word generator operating at 5 Gbit/s data rate, *IEEE Trans. Electron. Devices*, 29, 1094, 1982.
4. Baack, C., Integrierte Optik in der Kommunikationstechnik, *Nachrichtentech. Z.*, 37, 577, 1984.
5. Heydt, G., Teich, G., and Walf, G., 1.12 Gbit/s optical subscriber loop for transmission of high definition television signals, in Proc. 6th Int. Symp. Subscriber Loops Serv., Société des Electriciens des Electroniciens et des Radioelectriciens, Nice, 1984, 289.
6. Rein, H. M. et al., A time division multiplexer IC for bit rates up to 2 Gbit/s, *IEEE J. Solid-State Circuits*, SC-19, 306, 1984.
7. Teich, G., A channel selection module for gigabit line access, *Nachrichtentech. Arch.*, 6, 101, 1984.
8. Shikada, M., Emura, K., Fujita, S., Kitamura, M., Arai, M., Kondo, M., and Minemura, K., *Electron . Lett.*, 20, 164, 1984.

# INDEX

## A

Absorption
  coefficients of, 74—75
  losses due to, 12, 35, 199
Acquisition, 160—161
AlGaAs/GaAs, 77
Aluminum oxides, 114
Amplifier, see also High impedance type amplifier; Main amplifier; Preamplifier; Transimpedance type amplifier
  broadband, 90, 120, 126—133
  differential, 171—176
  noise power of, 134
  optical, 202—203
Amplitude comparator, 88
Amplitude regeneration, 4, 170—178
Analogue circuits, 88
APDs, see Avalanche photodiodes
Astigmatic behavior, 60
Attenuation, material, 4, 12, 17—19, 183, 188
Attenuation-limited systems, 8—9
Avalanche effect, 76, 80
Avalanche photodiodes (APDs), see also Germanium avalanche photodiode, 74
  III—V, 139—142, 183
  noise in, 80—84
  performance of, 76—78
  silicon (Si-APD), 82—83

## B

Backscattering, see also Optical feedback, 185
Band gap, 74, 78, 83—84
Bandwidth, 79—80, 185, 188
Baseline regeneration, 167—170
Beam waist, 56—57, 59—60
Beat frequency control 160—161
BER, see Bit error rate
BH, see Buried heterostructure laser
Binary systems, 5, 8
Bipolar signal, 154
Bipolar transistor
  in common base configuration, 120, 123—124, 128—129, 131, 136—138
  future development of, 112
  modulation circuit with, 34
  in multiplexing, 149
  as noise source, 142
  properties of, 96—105
Birefringence, 15, 17
Bit error rate (BER), 84, 183—185, 189
Bit rate, 196, 200
Bit rate-distance product, 4, 183, 185—188
Bonding, 95, 105, 114—115
Boron, doping with, 12, 14
Bridged integrator, 160

Brillouin scattering, 17
Buried heterostructure (BH) laser
  coupling in, 60, 63—64, 66—67
  dynamic properties of, 24—26, 30
  frequency modulation spectrum for, 32
  in Gb/sec systems, 184, 189
  relative intensity noise in, 39
  small signal equivalent circuit for, 32—33
  spectra of, 45, 47—48
Burst errors, 186, 188

## C

Capacitances, stray, 96
Capacitors, 91—93, 110
Carbon-film resistor, 93—95
Carrier spacing, 199
Cascode circuit, 131—133, 139
Ceramic materials, 114
Channeled substrate planar (CSP) laser
  coupling in, 61—62, 65—66, 185
  frequency modulation spectrum for, 32
  properties of, 23—24
  relative intensity noise in, 36—37
Chip capacitor, 91
Chip resistor, 93—94, 111
Chromatic dispersion, 4
  components of, 15
  low, fibers possessing, 186
  in 1.5-$\mu$m window, 18
  repeater spacing related to, 9
  of step-index fiber, 18—19
  transmission properties related to, 15—17
  zero points of, 17—20
Circuit design, 87—115
  active components, 95—109
  future developments, 112—113
  mounting technology, 113—115
  passive components, 91—96
  simulation and measurement of implemented circuits, 108—112
Cleaved coupled cavity ($C^3$) laser, 43, 46, 184
Clock content, 156, 158
Clocked comparator, 153, 170
Clock extraction, 161
Clock regeneration, see Timing regeneration
Coaxial cable (or line), 34—35, 95, 113, 168
Coherent light injection, 47—49
Common base configuration, 123—124, 128—129, 136—138
Comparators, 170—178
Competition noise, 39
Components, circuit
  active and passive, 91—113
  beyond Gb/sec range, 112
  commercially available, 88, 90—91
  monolithic integrated, 196

Compound cavity model, 65
Compound cavity mode region, 40—43
Computer-aided design, 77, 89—91
Confocal conditions, 64
Connector losses (or reflection), 67—68, 189, 199
Connectors, optical, 67—68, 185
"Cosine roll-off" spectrum, 6
Coupling
    efficiency of, 56—68
        fiber interconnection, 66—68
        fiber-photodiode coupling, 68—69
        laser-fiber coupling, 59—64
    elements for, 4, 67, 184, 196
    laser-fiber, 59—66
    losses in, 15, 56, 58, 62
Crystal, waveguides of, 114, 199
CSP, see Channeled substrate planar laser
Current gain, 76, 96—98, 101, 106, 108
Current switch, 33—34, 111
Cut-off wavelength, 14, 16—17
Cylindrical lens, 60

### D

Damping factor, 158
Dark current noise, 80—84, 134
DBR, see Distributed Bragg reflector laser
Delay line, 157
Depletion region, 77, 79
Detectors, optical, 73—84
DFB, see Distributed feedback laser
Digital circuits, 88, 90, 112—113
Dispersion, see Chromatic dispersion; Material
    dispersion; Waveguide dispersion
Dispersion-shifted fiber, 8—9
Displacement tolerance, 61—64
Distance-bandwidth product, 39
Distributed Bragg reflector (DBR) laser, 39, 44—
    46, 187, 201
Distributed feedback (DFB) laser, 43—46, 183—
    184, 187, 201
Doping, 12, 14, 100
Double-balanced mixer, 159
Drawn waveguide, 12
Dynamic single-mode (DSM) laser, 5, 43—49,
    189, 196, 201
Dynamic single-mode (DSM) state, 39, 43
Dynamic spectral width, 29

### E

Early effect, 102—103
Electronic-to-optical (EO) conversion, 196, 201
Emitter coupled logic integrated circuit (ECL-IC),
    149
Emitter degenerated circuit, 176—178
End face coupling method, 67
Equalization circuit, 4, 8, 162—166
Etching, selective, 61

European PCM hierarchy, see Pulse code modu-
    lation hierarchy
Excess noise factor, 81—83
External mirrors (reflectors), 46—49, 183
External resonator, 4, 41—43, 184—185, 187, 190

### F

Faraday isolator (or rotator), 69, 185
Far-field angle, 57
Feedback parameter, 65
FET, see Field effect transistor
Fiber photodiode coupling, 68—69
Fibers, see also Multimode fibers; Single-mode fi-
    bers
    attenuation in, 4, 183, 188
    coupling faces of, 61
    dispersion in, 8—9, 12—14, 183
    doping of, 12
    interconnection of, 66—68
    losses in, 12
    materials for, 12—14
    real, 17—20
    strongly guiding, 18
    weakly guiding, 17
Field distribution, 14—15, 56
Field effect transistor (FET)
    dual gate, 153
    modulation circuit with, 34
    properties of, 105—108, 112—113
    semiconductor materials for, 112
Filter
    Gaussian, 6, 8
    loop, 160
    optical, 198
    receiver, 4
    recursive, 162—166
    transversal, 154—155, 162—168, 181
Filtering, 166—167
Flicker noise, 99
Fluorine, doping with, 12
Frequency compensation, 144
Frequency control loop, 160—161
Frequency dividers, 90, 112
Frequency modulation, 30—32
Frequency response, 80
Fresnel reflections, 58—59, 67—68

### G

GaAlAs, 22, 37—38
GaAlAs-GaAs, 112
GaAlAsSb, 78, 83
GaAs, 75, 92, 112—113, 196—197
Gain-guided laser (GGL), 60
Gain guiding, 23
GaP, 75
Gas laser, 202
Gaussian beam, 56—60

Gaussian distribution, 15
Gaussian pulse, 5—8, 157
Gb/sec systems, 183—191, 196
Germanium, 12, 14, 75, 78
Germanium avalanche photodiode (Ge-APD),
   82—84
   in Gb/sec systems, 183—184
   properties of, 6
   sensitivity of receivers with, 139—142, 183
   signal and noise powers in, 134—135
GGL, see Gain-guided laser
Glass, waveguides of, 114, 199
Gluing, 66
Graded-index rod lens, 63
Groove coupled cavity laser, 43, 46
Group delay time, 17
Group index, 12

InGaAsP/InP laser, 46
Injection locking, 43, 47—49
Injection state model, 68
InP, 77—78, 112—113, 197
Integrated circuits, 112—113, 149, 157, 159
Integrated optics, 196—197, 201
Intensity noise, 35—40, 43
Intermodulation products, 28
Internal current gain, 76
Intersymbol interference, 185—186
Ionization coefficient, 76—77, 82

# H

## J

Half-angle of intensity, 57
HDTV, see High-definition TV
HEMT, see High electron mobility transistor
Heterodyne measuring process, 80
Heterojunction semiconductor, 22, 196
HgCdTe, 78, 83
High-definition TV (HDTV), 200
High electron mobility transistor (HEMT), 112—
   113
High impedance (HIT) amplifier, 120—121,
   140—142
   with cascade circuit, 131—133
   with MESFET input stage, 130—131, 133
   noise behavior of, 138—139
   noise minimization with, 142—143
   with series feedback input stage, 130, 133
High-speed circuit, 88—91
HIT, see High impedance type
Hybrid circuit, 88, 113
Hybrid optics, 197

Jitter, 158, 173, 178—181, 186
Josephson elements, 113

## L

$\lambda$-muldex, 201
$\lambda$-multiplex technology, 198—199, 201
Large signal behavior, 28—29, 99—105, 108, 111
Laser-fiber coupling, 59—66
Laser mode partition noise (LMPN), 4, 9, 18, 38,
   185
Lasers, see also Buried heterostructure laser;
       Channeled substrate planar laser
   classification of, 23
   fundamental properties of, 22—25
   intensity fluctuations of, 35—39
   modulation behavior of, 25—32
   modulation circuits for, 32—35
   optical feedback in, 39—43
   as optical transmitters, requirements for, 22
   types of
       cleaved coupled cavity ($C^3$), 43, 46, 184
       distributed Bragg reflector, 39, 44—46, 187,
          201
       distributed feedback, 43—46, 183—184, 187,
          201
       dynamic single-mode, 5, 43—49, 189, 196,
          201
       gain-guided, 60
       gas, 202
       groove coupled cavity, 43, 46
       index-guided, 23—24, 26, 59—60
       InGaAsP/InP, 46
       master, 47—49
       multimode, 4, 28, 41, 183
       single mode, 28, 187
       slave, 47—49
       stripe geometry, 37—38
       transverse junction stripe, 25, 32
       V-groove, 29—31, 39, 63—64
Leakage currents, 142—143
LED, see Light emitting diode
Lensed connector, 67
Lens transformation, 58
Light emitting diode (LED), 22

# I

IGL, see Index-guided laser
Impedances
   of high impedance type amplifier, 130—132
   input, of laser, 32—33
   with MESFETs, 106, 108
   of NE 644 microwave transistor, 121
   of transimpedance type amplifier, 127—129
Impurities, 12, 115
Index-guided laser (IGL), 23—24, 26, 59—60
Index guiding, 24, 60
Index matching fluid, 67—68
Inductances, 95, 105
InGaAs, 77, 83
InGaAs/InP, 83
InGaAsP, 22—23, 77—78, 83, 184
InGaAsP/InP, 78, 83

Light injection mode region, 40, 43
Line widening, 191
Lithium niobate, 197
LMPN, see Laser mode partition noise
Loop filter, 160
Losses
    connector, 67—68, 189, 199
    coupling, 15, 56, 58, 62
    microbending, 17
    optical, 12
    radiation, 12, 35
    scattering, 12
    splicing, 17—18

## M

Main amplifier, 88, 109—110, 143—145
Maritime cables, 202
Master laser, 47—49
Material dispersion, 12—17
    compensation of, 18, 186
    maximum bit rate and, 43
    for silica, 18—19
    zero point of, 13—14, 17
Metal semiconductor field effect transistor
        (MESFET), 89, 120
    leakage current from, 142—143
    in multiplexing, 149
    properties of, 105—109, 113
    as receiver stage, 124—126, 130—131, 138—
        139
Microbending losses, 17
Microlens, 60—62, 65, 184
Microstrip lines, 95, 113
Microwave technology, 88
Minimount®, 114—115
Mismatching, maximum, 126—127
MM, see Multimode
MMICs, see Monolithic microwave integrated cir-
    cuits
Modal noise, 4, 12, 17
Mode hopping, 30, 39—40, 43, 186, 188
Mode selection, 4, 66, 191
Modulation, laser, 4, 25—32, 191, 202
Modulation circuits, 32—35
Modulation noise, 191
Monolithic integration, 157, 159, 200
Monolithic microwave integrated circuits
        (MMICs), 112—113
Mounting technology, 113—115
Multilevel systems, 5
Multimode (MM) fibers, 12, 58
Multimode (MM) lasers, 4, 28, 41, 183
Multiplexers, 112, 148—153
Multiplication factor, 79, 82—84, 134—135

## N

Natural frequency, 158

Narrow line width (NLW) laser, 202
Network analysis, 89—90
Network analyzer, 91, 94, 97
NLW, see Narrow line width laser
Noise
    components, 4
    current, 81, 136
    driving forces, 35—36
    enhancement by reflection, 68, 189
    minimization of, 142—143
    power, 133—139
    side peaks, 185
    sources, 80—84, 136, 138
    types of
        competition, 39
        dark current, 80—84, 134
        flicker, 99
        intensity, 35—40, 43
        laser mode partition, 4, 9, 18, 38, 185
        modal, 4, 12, 17
        modulation, 191
        partition, 69, 185—186, 190
        relative intensity, 36—39
        Schottky, 99
        side mode, 36—37
        signal-dependent, 134
        thermal, 99, 134
Nonlinearities, 28
Nonlinear optical effects, 17
Normalized frequency, 14, 56
NPN transistor, 100, 102
Numerical aperture, 14, 56
Nyquist criterion, 156
Nyquist frequency, 139
Nyquist pulse, 5—8

## O

OEICs, see Optoelectronic integrated circuits
OERs, see Optoelectronic receivers
Optical detectors, 73—84
Optical feedback, 39—43, 58—59
    in fiber-photodiode coupling, 68—69
    in laser-fiber coupling, 65—67
    from optical connectors, 67—68, 185
    in practical systems, 190—191
    suppression of, 4, 196
Optical heterodyne systems, 17, 196—201
Optical isolator, 4, 69
Optical losses, 12
Optical sources
    dynamic properties
        laser modulation circuits, 32—36
        modulation behavior, 25—29
        spectral behavior, 29—31
    dynamic single-mode laser, 43—49
    intensity noise, 35—40
    optical feedback, 39—43
Optical-to-electronic (OE) conversion, 196, 201
Optical transmission systems, 4—6, 8, 22

Optoelectronic integrated circuits (OEICs), 196—197, 200—201, 203
Optoelectronic receivers (OERs), 6, 120—145
  amplifier concepts, 120—121
  main amplifier, 143—145
  noise behavior, 133—144
  structure, 126—133
Orthogonal modes, 17
Oscillators, 159

## P

Packaging, transistor, 105
Parasitic elements, 88
  method for determining, 92, 94
  in mounting, 113
  in oscillators, 159
  reduction of influence of, 126—127
Partition noise, 69, 185—186, 190
PCM, see Pulse code modulation
Performance, 183—192
Peltier effect cooling, 84
Penalty, system, 183, 190
Phase detector, 159—160
Phase-locked loop (PLL), 156, 158—162
Phosphorus, doping with, 12
Photocurrent noise, 80—83
Photodiode, see also Avalanche photodiodes
  avalanche gain of, 6
  coupling to fibers, 68—69
  equivalent circuit diagram of, 77
  in Gb/sec systems, 183—184
  materials for, 78
  in optoelectronic receivers, 120, 139—142
  PIN, 74—79, 139—141, 197
  properties of, 74, 79—84, 196
PLL, see Phase-locked loop
Plug connectors, 199
Polarization, 17
Postcursor equalization, 164—165
Power, maximum transmittable, 17
Power-coupling efficiency, 58
Power density spectrum, 190
Preamplifier, 88, 120
Preprocessing, 156—158
Propagation constant, 12
Pseudo-statistic signal, 190
Public networks, 199—203
Pulse code modulation (PCM) hierarchy, 148, 183
Pulse forms, 88—89
Pulse shaping, 153

## Q

Quadruple-clad (QC) fiber, 19—20
Quantized feedback equalization (QFE), 4—5, 88, 168—169
Quantum efficiency, 74—76, 79
Quartz (SiO₂), 4, 12—14, 18, 199

Quaternary semiconductor materials, 77, 83

## R

Radiation losses, 12, 35
Raman scattering, 17
Rate equations, 25—26, 41
Rayleigh scattering, 12, 14, 18, 59, 199
Receiver sensitivity, 139—142, 183
Reflection factor, 59
Reflections, far and near, 68—69
Refractive index
  effective, 15—16
  jumps, lightwave guidance by, 23—24
  of quartz, 12
  in real fibers, 17
  shifts in, 68
  of single mode fibers, 19—20
  temperature dependence of, 30—31
  wavelength dependence of, 12—13
Regeneration circuit, 153
Relative intensity noise (RIN), 36—39
Repeater spacing, 4—5, 7—9, 133
Residual dispersion, 4
Resistors, 93—95, 99, 111
Responsivity, 74—76
RIN, see Relative intensity noise

## S

SAM, see Separated absorption and multiplication
Sampler circuit, 153
Scattering losses, 12
Schmitt trigger, 155—156, 180
Schottky noise, 99
Selfoc® leases, 62—66, 184, 187
Self-pulsations, 43
Semiconductor materials, 22, 78
Separated absorption and multiplication (SAM) structures, 84
Series feedback stage, 120, 122—123
  of high impedance type amplifier, 130, 138
  in main amplifier, 144—145
  of transimpedance type amplifier, 127
Shunt feedback stage, 120—122
  of high impedance type amplifier, 131
  in main amplifier, 143—144
  of transimpedance type amplifier, 127—128, 136, 138
Side modes
  fluctuations in, noise due to, 36—37
  intensity in, 43, 45, 47—49
  suppression of, 29, 41—43, 48—49, 190
Signal
  amplification of, 4
  distortion of, 162, 186
  generation of, 148—155
  regeneration of, 155—181

clock, 155—162
waveform, see also Waveform regeneration,
    162—181
Signal-dependent noise power, 134
Signal-to-noise ratio (SNR)
    deterioration due to fiber dispersion, 38
    of a Ge-APD, 84
    increase due to noise component compensation,
        37
    of input signal at comparator, 171
    optimization rules for, 5
    at optoelectronic receiver output, 133—134,
        143
Silicon, 75, 78, 112, 196
Silicon dioxide (SiO₂), see Quartz
Single-mode (SM) fibers
    channel allocation in, 198—199
    coupling efficiency in, 56—58
    design of, 17—18
    field distribution in, 14—15, 56
    in Gb/sec systems, 184
    Ge-doped, losses in, 12—14
    necessity for, 4—5, 12, 196
    optical feedback in, 58—59
    polarization-maintaining, 17
    properties of fiber materials, 12—14
    reflection in, 58
    refractive index profiles of, 19
    spot size of, 56—58
    step-index, 15, 18—19
    in subscriber field, use of, 199
    transmission properties of, 6, 15—17, 199
    with triangular index profile, 19
    as waveguides, 14—15
Single-mode (SM) lasers, 28, 187
Slave laser, 47—49
SM, see Single-mode
Small signal behavior, 26—28, 33, 43
    of bipolar transistors, 96—99
    example of, 108—110
    of MESFETs, 106—108
SNR, see Signal-to-noise ratio
Splicing, 66, 199
    losses due to, 17—18
Spot size, 17—19, 56—58
Step-index fibers, 15, 18—19
Step recovery diodes, 113
Stripe geometry laser, 37—38
Stripline cables, 113
Submodes, 67—68
Subscriber lines, 199—201
Sweeping, 160
Switching times, transistor, 96—97
Symmetry method, 156—157
Synthetic substrates, 114
System margin, 185, 190
System performance, 183—192

**T**

Tapers, 61—66, 184—185, 190
Telephone service, 200, 202
Temperature
    dark current, effect on, 84
    mode hopping related to, 30, 43
    polarization at end of fiber and, 17
    refractive index dependence on, 30—31, 42—43
    spectral changes related to, 29—30
Ternary semiconductor materials, 77, 83
Thermal noise, 99, 134
Thick and thin film technologies, 113—114, 144
III-V materials, 22, 78, 196—197
Threshold method, 156
Timing regeneration, 4—5, 88, 155—162, 178—
        180
TIT, see Transimpedance type amplifier
TJS, see Transverse junction stripe
Transient response, 121—122
Transimpedance type (TIT) amplifier, 120—121,
        140—142
    with common base input stage, 128—129, 133
    in Gb/sec systems, 183—184
    noise behavior of, 136—138
    with shunt feedback input stage, 127—128,
        133, 138
Transistor, see also Field effect transistor; Metal
        semiconductor field effect transistor
    chips, 106, 111—112
    high electron mobility, 112—113
    modeling of, 96—105
    NPN, 100, 102
Transistor stages, 120—127
Transit frequency, 96, 98, 126
Transmission lines, 95
Transmission methods, 5—9
Transmission rates, see also Bit rates, 4
Transmission systems, long-haul, 18
Transversal confinement, 23
Transversal filter, 154—155, 162—168, 181
Transverse junction stripe (TJS) lasers, 25, 32

**V**

V-groove laser, 29—31, 39, 63—64
Video telephone transmission, 200, 202
Voltage-controlled oscillator (VCO), 158

**W**

Waveform regeneration, 162—181
    amplitude regeneration, 170—179
    baseline regeneration, 167—170
    complete regeneration circuit, 180—181
    equalization, 162—166
    filtering, 164—166
    timing, 179—180
Waveguide

dimensioning of, 7, 14
dispersion of, 15—16, 18, 186
drawn, 12
field distribution in, 14—15
material for, 12, 14, 199
Wavelength division multiplex (WDM) systems,
186, 198
Wavelength shifting, 18

Word generators, 112
Working distance, 61

## Z

Zero-dispersion wavelength, 13—14, 17—20, 43,
183